Chaos and Noise in Biology and Medicine

Istituto Italiano per gli Studi Filosofici

Series on Biophysics and Biocybernetics

Coordinating Editor: C. Taddei-Ferretti

Vol. 1: Biophysics of Photoreception: Molecular and Phototransductive Events
edited by: C. Taddei-Ferretti

Vol. 2: Biocybernetics of Vision: Integrative Mechanisms and Cognitive Processes
edited by: C. Taddei-Ferretti

Vol. 3: High-Dilution Effects on Cells and Integrated Systems
edited by: C. Taddei-Ferretti and P. Marotta

Vol. 4: Macromolecular Interplay in Brain Associative Mechanisms
edited by: A. Neugebauer

Vol. 5: From Structure to Information in Sensory Systems
edited by: C. Taddei-Ferretti and C. Musio

Vol. 6: Downward Processes in the Perception Representation Mechanisms
edited by: C. Taddei-Ferretti and C. Musio

Forthcoming volumes:

Vol. 8: Neuronal Bases and Psychological Aspects of Consciousness
edited by: C. Taddei-Ferretti and C. Musio

Vol. 9: Neuronal Coding of Perceptual Systems
edited by: W. Backhaus

Vol. 10: Emotions, Qualia and Consciousness
edited by: A. Kaszniak

ISTITUTO ITALIANO PER GLI STUDI FILOSOFICI
SERIES ON BIOPHYSICS AND BIOCYBERNETICS
Vol. 7 – Biophysics

Chaos and Noise in Biology and Medicine

Proceedings of the International School of Biophysics
Casamicciola, Napoli, Italy, 19–24 May 1997

Edited by

M. Barbi
S. Chillemi

Istituto di Biofisica, CNR, Pisa, Italy

World Scientific
Singapore • New Jersey • London • Hong Kong

Published by
World Scientific Publishing Co. Pte. Ltd.
P O Box 128, Farrer Road, Singapore 912805
USA office: Suite 1B, 1060 Main Street, River Edge, NJ 07661
UK office: 57 Shelton Street, Covent Garden, London WC2H 9HE

British Library Cataloguing-in-Publication Data
A catalogue record for this book is available from the British Library.

CHAOS AND NOISE IN BIOLOGY AND MEDICINE

Copyright © 1998 by World Scientific Publishing Co. Pte. Ltd.

All rights reserved. This book, or parts thereof, may not be reproduced in any form or by any means, electronic or mechanical, including photocopying, recording or any information storage and retrieval system now known or to be invented, without written permission from the Publisher.

For photocopying of material in this volume, please pay a copying fee through the Copyright Clearance Center, Inc., 222 Rosewood Drive, Danvers, MA 01923, USA. In this case permission to photocopy is not required from the publisher.

ISBN 981-02-3600-X

This book is printed on acid-free paper.

Printed in Singapore by Uto-Print

PREFACE

This is the seventh volume of the Series on Biophysics and Biocybernetics, promoted by the Istituto Italiano per gli Studi Filosofici. It appears as the Proceedings of the Course on "Chaos and Noise in Biology and Medicine" of the International School of Biophysics, which was inaugurated at Palazzo Serra di Cassano, Naples, Italy, and was held at Casamicciola on the isle of Ischia, Italy, on May 5-10, 1997, under the direction of this volume's Editors.

The School, which dealt with borderline topics from physics, mathematics, engineering, biology and medicine, was promoted and supported by the Istituto Italiano per gli Studi Filosofici and was organized by the Istituto di Biofisica of CNR (Pisa).

We would like to thank the members of the School Advisory Board, namely, V. Anischenko (Russia), R. Balocchi (Italy), J. Collins (USA), A.V. Holden (UK), J. Kurths (Germany), M. Spano (USA) and C. Taddei-Ferretti (Italy), for their fruitful advice and suggestions as well as for their helpful cooperation. We would also acknowledge the partial financial support of the Gruppo Nazionale di Cibernetica e Biofisica of the Italian Consiglio Nazionale delle Ricerche (CNR), the Comitato Nazionale per la Biologia e Medicina (CNR) and the Comitato Nazionale per la Fisica (CNR). Furthermore, only due to the help of the Istituto Italiano per gli Studi Filosofici, it has been possible to award grants to several participants, especially those coming from deserving countries.

We also wish to thank all the scientists who gave one or two lectures and, with their discussions, contributed to create the charming atmosphere of a highly stimulating scientific milieu. Unfortunately, some of the main lecturers sent us only one paper and two of them neither that. Instead, all the participants who gave a talk, provided us with the related paper.

Finally, we thank the members of the local organizing committee: A. Cotugno, A. Di Garbo, C. Musio, C. Neri, S. Santillo, of the Istituto di Biofisica (CNR) and Istituto di Cibernetica (CNR); without their work the School could not have been realized, nor its cordial atmosphere obtained. Moreover, we acknowledge the precious support of N. Aprile, of the Istituto Italiano per gli Studi Filosofici, C. Petrongolo and G. Tocchini of the Istituto di Biofisica (CNR), of the administrative staff.

The beauty of the isle of Ischia and the courtesy of the staff of the Hotel Gran Paradiso at Casamicciola, where the School was held, completed the pleasantness of the environment.

Michele Barbi and Santi Chillemi

CONTENTS

Preface   v

INTRODUCTORY LECTURE

Stochastic resonance in sensory biology: from single neurons to the human visual cortex   3
    F. Moss, X. Pei, E. Simonotto and L. Wilkens (St. Louis, MO, USA)

NONLINEARITY AND CHAOS IN NEURONAL AND CARDIAC DYNAMICS

Transmission of a chaotic signal by a synaptic-like coupling   25
    S. Chillemi, M. Barbi and A. Di Garbo (Pisa, Italy)

System reconstruction and analysis using interspike intervals   36
    R. Castro and T. Sauer (Fairfax, VA, USA)

Surrogate data for neural activity   51
    D. Petracchi (Pisa, Italy)

Low dimensional dynamics in sensory biology   60
    F. Moss, X. Pei, K. Dolan, H. Braun, M. Huber and M. Dewald
    (St. Louis, MO, USA and Marburg, Germany)

Fractal analysis in the heart   77
    R. Balocchi, C. Michelassi, C. Carpeggiani M. Castellari and
    M. G. Trivella (Pisa, Italy)

Nonlinearity of normal sinus rhythm   86
    M. Barbi, S. Chillemi, A. Di Garbo, R. Balocchi, C. Carpeggiani,
    M. Emdin and C. Michelassi (Pisa, Italy)

CHAOS CONTROL AND ITS APPLICATIONS

Adaptive strategies for recognition and control of chaos 99
    F. T. Arecchi and S. Boccaletti (Firenze, Italy)

Controlling chaos with applications to nonlinear digital communication 114
    C. Grebogi, Y.-C. Lai and S. Hayes
    (Potsdam, Germany and College Park, MD, USA)

Controlling and maintaining chaos in biological systems 134
    M. L. Spano, V. In and W. L. Ditto (Bethesda, MD, USA)

SPATIO-TEMPORAL DYNAMICS

Spatio-temporal irregularity of propagating electrical activity in models of cardiac tissue: the genesis of flutter and fibrillation 147
    A. V. Holden (Leeds, U.K.)

Dynamical control of biological rhythms 162
    D. J. Christini and J. J. Collins (Boston, MA, USA)

The dynamics of coupled active oscillators 176
    L. Fronzoni (Pisa, Italy)

Approaches to defibrillation: the control of re-entrant activity, and the annihilation of propagating waves of excitation in cardiac tissue 195
    A. V. Holden (Leeds, U.K.)

NOISE AND STOCHASTIC RESONANCE

Chaos and noise in excitable systems 213
    R. Castro and T. Sauer (Fairfax, VA, USA)

COMMUNICATIONS

Detection of small-conductance Cl- channels by wavelet transform 225
    M. Nobile, L. Lagostena, R. Fioravanti (Genova, Italy)
    and S. Ferroni (Bologna, Italy)

Brownian motion and the ability to detect weak auditory signals   230
    I. C. Gebeshuber, A. Mladenka, F. Rattay and
    W. A. Svrcek-Seiler (Wien, Austria)

Delayed luminescence in biophysical investigations   237
    A. Scordino, A. Triglia, F. Musumeci (Catania, Italy)
    and R. Van Wijk (Utrecht, The Netherlands)

Painlevé property of ODEs and time series analysis   244
    A. Di Garbo, M. Barbi and S. Chillemi (Pisa, Italy)

Evaluation of self-similarity and regularity patterns in heart rate
variability time series   249
    M. G. Signorini (Milano, Italy)

On the characterization of atrial fibrillation using wavelets and
correlation dimension   257
    A. Casaleggio and R. Fioravanti (Genova, Italy)

Dynamical system modelling and control of a cardiac arrhythmia   262
    T. Nomura, Y. Saeki and S. Sato (Osaka, Japan)

Investigation of passive electrical properties in the right atrium   271
    A. Grigaliūnas and R. Veteikis (Kaunas, Lithuania)

On the stochasticity of neuronal firing patterns: the local stability of
limit cycles   275
    F. Ali and M. Menzinger (Toronto, Canada)

Oscillations in continuous-time graded-response neural networks with delay   283
    K. Pakdaman, C. P. Malta, C. Grotta-Ragazzo, O. Arino
    and J. F. Vibert (Paris and Pau, France, Osaka, Japan and
    Sao Paulo, Brazil)

Potentiality of complex behaviour in single neurons   288
    L. Andrey (Prague, Czech Republic)

Determinism versus stochasticity in biological interspike interval
time series 294
   N. Stollenwerk (Jülich, Germany)

Synchronization and chaos in networks consisting of interacting
bistable elements 301
   V. Chinarov (Kiev, Ukraine)

Stochastic model of population dynamics 305
   F. De Pasquale and B. Spagnolo (Roma and Palermo, Italy)

Analyzing the multifractal structure of DNA nucleotide sequences 315
   J. M. Gutierrez, A. Iglesias, M. A. Rodriguez (Santander, Spain)
   J. D. Burgos, C. M. Estevez and P. Moreno (Bogotà, Colombia)

Chaotic firing of cortical neurons upon inhibitory periodic stimulation 320
   R. Stoop, K. Schindler and P. Goodman (Zürich, Switzerland)

PARTICIPANTS

List of the participants 331

INTRODUCTORY LECTURE

STOCHASTIC RESONANCE IN SENSORY BIOLOGY: FROM SINGLE NEURONS TO THE HUMAN VISUAL CORTEX

FRANK MOSS, XING PEI, ENRICO SIMONOTTO and LON WILKENS
Center for Neurodynamics, University of Missouri at St. Louis
Saint Louis, MO 63121, USA
E-mail: mossf@umslvma.umsl.edu

ABSTRACT

We describe a method, called *stochastic resonance* (SR), whereby random processes in noisy systems can aid in the detection or transmission of weak information carrying signals. Though SR has a long history of applications in physics, particular emphasis is placed on its occurrence in biological systems. The method is based on the statistical detection of threshold crossing events, which is the simplest picture of how a neuron functions. Examples are given from neural recordings obtained from single sensory neurons in the tailfan of the crayfish and from the photoreceptor cell embedded in its 6th ganglion. The use of SR as a tool in visual perception is illustrated.

1. Introduction

Stochastic resonance (SR) is a term, which describes dynamical behavior resulting from the nonlinear interaction of noise with a weak signal. It was first proposed as a process for describing the observed periodicities in the recurrence of the earth's ice ages (Benzi, *et al.,* 1981, Nicolis, 1982, 1993). Since 1982, it has found wide application in physics (Jung, 1993; Moss, 1994a; Hibbs, *et al.,* 1995) and in biology (Wiesenfeld and Moss, 1995; Moss, *et al.,* 1994b). In particular it has been demonstrated as a sensory process in the hydrodynamically sensitive mechanoreceptors of the crayfish (Douglass, *et al.,* 1993; Pei, *et al.,* 1996) and the air motion sensitive receptors of the cricket (Levin and Miller, 1996). Both animals use their motion detectors to escape predators. Both animals are of the same phylum but separated in time on the evolutionary scale by more than 100 million years. The cricket is a relatively modern land animal, whereas the ancestors of the crayfish, equipped with similar water motion sensing apparatus, were swimming in the Cambrian seas. This suggests that SR may have played a role in sensory biology since the earliest evolutionary responses to the appearance of predators.

1.1. SR in Bistable Systems

The original applications of SR were to systems (mostly physical) with bistable energy potentials (Lanzara, *et al.*, 1997) for example, as shown in Fig. 1.

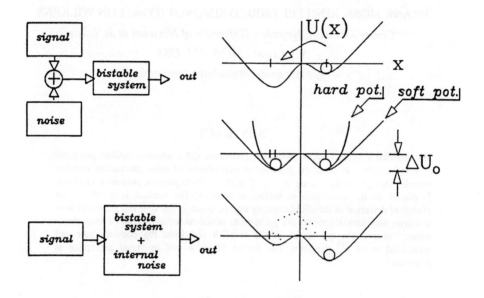

Fig. 1. A bistable potential with weak periodic modulation which alternately raises and lowers the potential wells relative to the barrier, ΔU_0. The shape of the potential can vary, for example, as shown by the "hard" and "soft" shapes shown. In the absence of added noise, the state point, symbolized by the balls, cannot switch wells. With added noise, say Gaussian, there is a non-zero, periodically modulated, probability that the ball will switch wells as indicated by the arrows. Information is detected or transmitted only by switching events. Thus noise is necessary for the transmission or detection of information. As shown on the left, the noise can either be internal to the system or external arising within the environment.

The switching events carry a surprising amount of information about the weak, periodic signal. This can be easily seen by considering the Kramers rate which determines the mean switching rate, R, and thus the information carrying capability:

$$K = \nu_0 \exp\left[-\frac{\Delta U_0}{<\xi^2>}\right] \tag{1}$$

where ΔU_0 is the barrier height, $\langle\xi^2\rangle$ is the noise intensity and ν_0 is the barrier attempt frequency. In the simplest view, the periodic modulation is introduced as,

$$\Delta U_0 \rightarrow \Delta U_0 + \varepsilon \sin \omega t \qquad (2)$$

where the weak signal amplitude $\varepsilon \ll \Delta U_0$. Thus the mean barrier crossing rate, and hence the mean information transmission rate, become *exponentially sensitive* to the weak periodic signal.

Various information measures can be applied to the time series of barrier crossing events. The most often used one is the signal-to-noise ratio (SNR). This is usually obtained from the averaged power spectrum of the crossing event time series. Example time series and power spectrum are shown in Fig. 2. The time series are shown in the upper two panels and the power spectrum at the bottom. Note that the time series in the middle panel has been two-state filtered, since the essential information about the underlying weak signal is contained in the barrier crossings alone. The signal features in the power spectrum show up at the odd harmonics only, because the data were obtained for a symmetric bistable potential, in this case the standard quartic potential: $U = -x^2/2 + x^4/4$. The SNR, as defined in the figure caption, is shown in Fig. 3 as a function of the noise intensity. Note the signature of SR is the maximum in the SNR curve located at an optimal value of the noise intensity. The theory of periodically modulated,stochastic, nonlinear dynamical systems, including the theory of SR, has been developed in great detail (Jung, 1993).

1.2. Non-dynamical SR

There is a simpler paradigm of SR. It occurs in purely statistical, that is, non-dynamical systems (Gingl, *et al.*, 1995; Moss, *et al.*, 1994b). The minimum necessary requirements for a systems to exhibit non-dynamical SR are 1) a threshold, 2) a subthreshold signal, and 3) noise. In this case, information about the subthreshold signal is conveyed by threshold crossing events. Noise is added to the subthreshold signal, so that the signal plus the noise crosses the threshold at some mean rate. These threshold crossings are marked by a temporal sequence of identical pulses, which form the system output. Each pulse carries only one bit of information: the time at which a threshold crossing occurred. Assuming the noise to be Gaussian and colored, that is, of limited bandwidth and exponentially correlated, the probability of a threshold crossing is a purely statistical quantity, which depends only on the noise intensity and the distance between the mean of the noise distribution and the threshold.

A simple approximate theory, which has the virtue of being quite transparent, has been developed (Gingl, *et al.*, 1995; Moss, *et al.*, 1994b). We quote only the final formula for the SNR here:

$$SNR = \log_{10}\left[\frac{2f_0 \Delta_0^2 \varepsilon^2}{\sigma^4 \sqrt{3}}\right] \exp\left[-\frac{\Delta_0^2}{2\sigma^2}\right], \qquad (3)$$

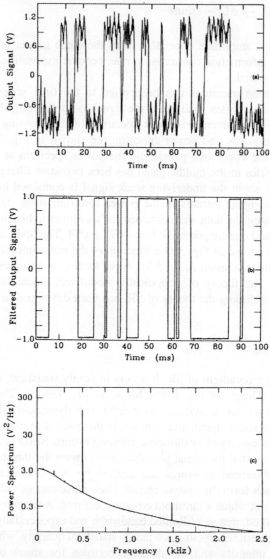

Fig. 2. Time series and power spectrum. (a) a time series of the actual state point showing noisy barrier crossings and the effect of noise on the state point within the wells. (b) a similar time series which has been "two-state filtered", that is, the noise within wells filtered out. (c) the power spectrum of the time series in (b) showing the Lorentzian broadband noise background with signal features which show as sharp peaks at the fundamental and third harmonic of the signal frequency. The SNR is defined as the ratio of the area under the fundamental signal feature to the background noise intensity in a 1 Hz bandwidth centered on the fundamental frequency.

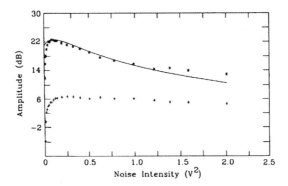

Fig. 3. The SNR as a function of the noise intensity. The defining formula is a standard definition from communications theory: $SNR = 10\ log(A/N(\omega))$ in decibels (dB). A is the area under the signal feature and $N(\omega)$ is the noise intensity at the signal frequency multiplied by 1 Hz. The asterisks are measurements made on an electronic circuit model of the standard quartic. The crosses are the noise intensity N, measured from the power spectrum. The theory (see, for example, Jung, 1993) is shown by the solid curve.

where f_0 is the bandwidth of the noise (it cannot be white, since in that case the threshold crossing rate is infinite), Δ_0 is the distance between the threshold and the signal mean, and σ is the standard deviation of the noise, which is assumed to be Gaussian. One can note that occurrence of σ in the denominators of both the prefactor and the exponent leads to a maximum SNR at an optimal noise intensity.

The non-dynamical model of SR is, in fact, also one of the simplest models of the neuron. The subthreshold signal represents some analog information on the membrane potential. Whenever the membrane potential crosses the threshold, the neuron "fires", that is generates a propagating action potential in the form of a standard shaped pulse. The output of a neuron is thus a temporal sequence of action potentials which mark the threshold crossing times of the signal plus noise on the membrane potential. Moreover, it is characteristic that neurons are noisy, that is the membrane potential is noisy even without a signal. Thus neurons characteristically fire spontaneously at random times — a process which is called the "internal" noise. In view of these observations, it is not surprising that SR has been found in biological sensory systems.

In order to illustrate the aforementioned ideas, we show in Fig. 5, the SR characteristic of an electronic FitzHugh-Nagumo neuron model (Moss, et al., 1993). This model is a reduced version of the Hodgkin-Huxley model. It is characterized by a fast limit cycle (the action potential) followed by a slow recovery (the refractory period, during which it is not possible for the neuron to fire). The model shows a fixed point below a threshold which marks the onset of limit cycle (action potential) generation. The model is operated in the subthreshold regime

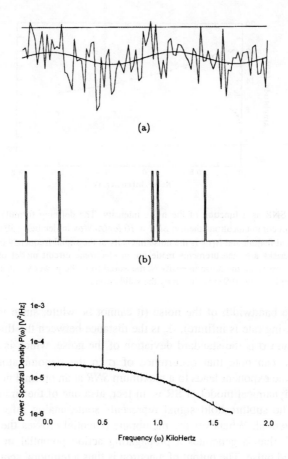

Fig. 4. (a) A simulation of a threshold (straight line), a subthreshold signal (sinusoid) plus noise. (b) Threshold crossings are marked by the sequence of standard pulses. (c) The power spectrum showing signal features at the fundamental and second harmonic frequencies.

where a weak signal plus Gaussian colored noise is added. The model responds with a more-or-less random sequence of action potentials, which, however, carry information about the subthreshold signal. Thus SR can be observed, as shown by the *SNR* measurements in Fig. 5.

This serves to illustrate the utility of the non-dynamical approach. The FitzHugh-Nagumo model is a dynamical system, but because the action potentials occur on a much faster time scale than the signal and noise, the system is well approximated by the purely statistical theory.

In the following, we will show how SR was discovered in the crayfish sensory system, how it persists into the outer levels of the animal's central nervous system

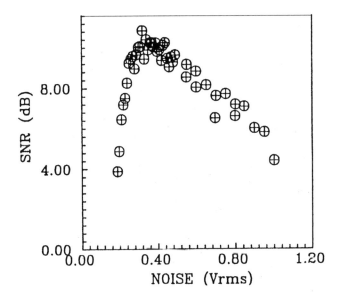

Fig. 5. The SNR versus noise intensity measured on an electronic FitzHugh-Nagumo model neuron.

and how these findings were verified and extended in experiments with the cricket. Finally, we discuss an application of non-dynamical SR to psychophysical measurements in the human visual system.

2. Crayfish SR

Since the earliest days of neurobiology, the crayfish has served as an animal for fruitful study (Huxley, 1880). The part we are interested in is the tailfan with its array of hair receptors and the 6th, or terminal, ganglion with its pair of photoreceptor cells (Hermann and Olsen, 1967). The preparation is shown in Fig. 6.

This preparation is excised from the animal and the nerve roots and ganglion exposed surgically. It is then mounted in a Ringer solution natural to the fluids surrounding the neurons of this animal. Neural impulses can be studied at the sensory neurons as shown by the lower electrode-amplifier combination, or on the photoreceptor output neuron (Wilkens, 1988) as shown by the upper combination. The hairs are stimulated by relative fluid motions in the direction shown. These motions are typically sinusoidal of amplitude 10 to 100 nanometers at frequencies from 5 to 100 Hz and velocities 100 to 1000 microns per second. The responses of a photoreceptor cell can be studied in the presence of both hydrodynamic stimula-

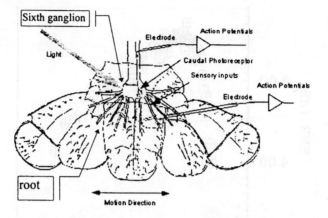

Fig. 6. The tailfan with approximately 250 long hairs specialized for sensing approximately periodic water motions. The hairs are connected to the interneurons within the ganglion by sensory neurons collected into nine nerve roots. The some interneurons are also connected to the two photoreceptor cells whose axons go upwards to the higher nervous system.

tion and steady light intensity on the photoreceptive area. The experimental protocols necessary for this research have been well-developed (Wilkens and Douglass, 1994).

2.1. SR in the Crayfish Sensory Neuron.

Recordings from the lower electrode, attached to a sensory neuron, show only noise plus evidence of the periodic stimulus. Using external, or environmental, noise plus a weak signal, SR can be observed at this recording location (Douglass, et al., 1993). We choose a sensory neuron with evidence of very low internal noise in order to clearly observe the effects of the external noise. Example data are shown by the squares in Fig. 7, where the theory is shown by the solid curve and a FitzHugh-Nagumo simulation is shown by the diamonds.

This experiment demonstrated SR in the crayfish sensory neuron using environmental noise as the random source. It suggests that the hair receptor system is specialized for early predator detection in the hydrodynamically noisy environment (rapidly flowing streams with turbulent water) where the animal is found. Moreover, the signals best detected are sinusoidal in the frequency range around 10 Hz. This is the frequency characteristic of the water motions induced by swimming fish (the predators). Periodic water motions from the swimming fish travel faster than the fish.

A major question concerning the internal noise of sensory (and other) neurons

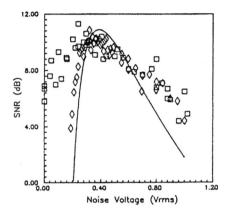

Fig. 7. Crayfish sensory SR is shown by the squares, compared to theory (solid line) and FitzHugh-Nagumo simulation (diamonds). The noise voltages on the horizontal axis correspond to random fluid motions in the range 25 to 200 nm.

remains open. Virtually all neurons are noisy (Moore, Perkel and Segundo, 1966; Adey, 1970), and some are much more noisier than can be accounted for by equilibrium statistical processes (Denk and Webb, 1992). If this noise is not useful for some process, one would think that evolutionary processes should have long since eliminated or minimized it. SR is a possible answer to this question. In order to further explore this topic experimentally, it will be necessary to manipulate the internal noise of neurons in a controlled way, that is, without affecting other properties of the neuron. Experiments designed to seek SR in the crayfish sensory neurons by controlling the internal noise by means of the temperature of the preparation have been performed. While interesting results were obtained, also with animals acclimated for long times at different temperatures, SR was not observed (Pantazelou, *et al.,* 1995).

2.2. SR in the Crayfish Photoreceptor Cell

There is another way to control the internal noise of a neuron which has resulted in the observation of SR. The light intensity falling on a photoreceptive area embedded in the 6th ganglion mediates the internal noise of that neuron. This has been shown by recordings from the photoreceptor output neuron at the location shown by the upper electrode in Fig. 6, where SR was observed as the light intensity was varied (Pei, *et al.,* 1996) Neural output at this location is determined by the complex (and largely not understood) computational processes occurring within the ganglion. Thus we can observe low dimensional dynamical behavior and possibly chaos, in any case, complex dynamics on the photoreceptor output axon (Moss, et al, in this volume). In these experiments, we applied hydrodynamic stimuli as usual, but accumulated data also in the presence of a steady light on the photoreceptor.

Example data are shown in Fig. 8, where the SNR is plotted against light intensity for two amplitudes of the stimulus.

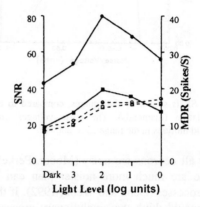

Fig. 8. The SNR versus the light intensity for two amplitudes of the 10 Hz sinusoidal, hydrodynamic stimulus. Solid squares: 0.26 microns; solid circles, 0.42 microns. The open symbols show the mean spike rate, or maintained discharge rate (right scale).

The experimental results outlined above have been confirmed in a recent experiment on SR in the cricket cercal system (Levin and Miller, 1996). The cricket is an animal of the same phylum as the crayfish, but its origins are much more recent on the evolutionary time scale. Nevertheless, the cricket makes use of a very similar predator avoidance system. The cricket has two rear appendages which are covered with hair mechanoreceptors. Each receptor is connected to a set of interneurons in a terminal ganglion very much like that of the crayfish. The hairs respond to air motions and have characteristic frequencies in the range 80 to 150 Hz. (By contrast, the crayfish mechanoreceptive system responds to hydrodynamic stimuli in the frequency range 8 to 25 Hz) Natural predators of the cricket are a certain species of wasp and, or course, birds. The wing beats of these animals lies within the frequency range characteristic of the cricket cercal system.

The SR experiment on the cricket was designed with many improvements. For example, modern information transfer measures rather than the simple SNR measure were used. The experiment was performed in a similar overall way: a weak periodic stimulus was applied to the appendages in the form of air motions to which random motions (the noise) were added. The results showed an optimal external noise intensity for which the transinformation from stimulus to interneurons in the ganglion was a maximum. Moreover, the results showed that the addition of external noise to a weak periodic signal could also improve the neural action potential timing precision.

3. SR as a Tool for Quantifying Human Visual Processes

How does the human brain process noisy visual information. We are surprisingly adept at it. We can observe a faint outline in swirling fog and detect and interpret it presently much quicker and more accurately than can the best computers with the most advanced pattern recognition software. What is going on? The original motivation for the experiments described in this section was to replace the data analysis software and the computers invariably used in all preceding SR experiments to analyze neural recordings with a human interpretation. It was easy to show that noise enhances the information content in some neural recording in the nervous system, but do the animals really make use of the enhanced information? These experiments were designed to make a step in that direction by addressing the question: can humans interpret noise enhanced visual information?

The experiments have been described in detail elsewhere (Simonotto, *et al.,* 1997) and so will only briefly be outlined here. A picture was digitized on a 256 level gray scale and displayed on a computer monitor as an image of 256 by 256 pixels. The picture was then sunk beneath a threshold. We now have a subthreshold visual "signal". The gray values of all pixels lie beneath the threshold so the image on the monitor is blank. Noise is then added to this subthreshold image by choosing a number from a zero-mean Gaussian distribution and adding this to the image pixel-by-pixel. The noise in each pixel is uncorrelated with the noise in every other pixel. Some pixels now contain gray levels above the threshold, and these are painted completely black. The remaining pixels (with gray levels below the threshold) are painted white). If the noise added is too small, too few pixels contain information about the image and the resulting picture is difficult to interpret. If the noise is too large, there is too much randomness in the picture and it is again difficult to interpret. An optimal noise results in the most interpretable picture. An example is shown in Fig. 9. Note that one can only read the time in the middle panel, which represents the optimal noise added.

The effects on the human visual system are dynamical, but unfortunately the static printed picture on this page cannot convey this. What happens is that when the noise in every pixel becomes a function of time, and the picture is then presented as a motion picture, the enhancement power of the added noise is greatly improved. For an interactive animation please go to the site on the World Wide Web designated in the caption of Fig. 9. In the dynamical presentation, the noise correlation time also has an effect which can be observed at the Web site.

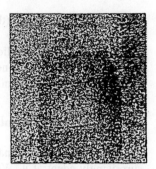

Fig. 9. Three noise contaminated, threshold filtered pictures of Big Ben. The noise added to the subthreshold image is optimal in the center panel. For an interactive, animated example of the effects of dynamic noise and noise of different correlation times, see the URL http://www.ge.infm.it/~simon/sr.html

In the psychophysical experiment, subjects were presented with series of pictures contaminated with noise of varying intensity and/or correlation time. The image used was not a building but instead a standard pattern used in visual psychophysics experiments. Subjects were asked to identify a fine detail of the pattern. Their perceptive threshold A_{th} was measured. Example data from one subject are shown in Fig. 10. Note that the perceptive threshold registered by the subject is minimum when the information transmitted by the picture to the visual cortex is maximum. Thus the characteristic signature of SR appears in this experiment as a minimum in the perceptive threshold. The standard equation describing SR, Eq. (3), was rewritten as a signal amplitude equation, inverted, and solved for a multiplicative constant A_{th}, which has the dimensions of a signal power. A number of constants in the prefactor are lumped into a single constant K, which was used as the only adjustable parameter in fitting the psychophysical data with this theory. The equation is shown on Fig. 10. Note that the fit is not excellent, but is surprisingly good in view of the implicit assumptions. In particular Eq. (3) was derived for a single element stochastic resonator. But brain processes are complex so there is no reason to suppose that a single element equation would describe the data. Moreover, Eq. (3) was derived from the Fourier transform of a time series. There is no reason to believe that the brain computes Fourier transforms. But the agreement is remarkable none-the-less and leads one to speculate on visual perceptive and brain processes.

Note that the fitting constant K applies to a single individual subject and is a measure of his or her ability to detect and interpret fine detail in a noise contaminated visual picture. This ability varies considerably from subject-to-subject. We have tested to date 16 subjects, with results shown in Fig. 11. Each

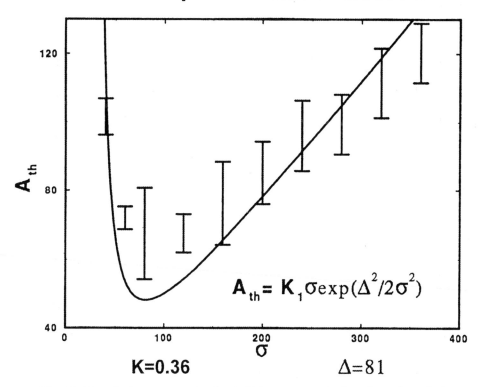

Fig. 10. Example data on perceptive threshold A_{th}, versus noise intensity σ, for a single subject. Note the minimum in perceptive threshold at the optimal noise intensity $\sigma \cong 82$. The equation shown (solid curve) was fit to the data using a single fitting constant K, the value of which is shown. For all subjects the threshold was held constant as shown. In a single session, the subject was presented with the same image but with one of 10 different noise intensities in a randomized sequence. Each noise intensity was (randomly) presented 10 times for a total of 100 presentations. At each noise intensity the error bar represents twice the standard deviation of the 10 subject reported perceptive thresholds.

subject was tested three times, as shown by the groupings of three symbols, in three sessions separated by at least one day and often as much as one week. Note the repeatability for an individual subject over the three sessions. Moreover, we recalled three of the subjects after approximately one year with the results shown by the diamonds. These results indicate that our experiment produces robust and repeatable results. All subjects were healthy young people between the ages of 18 and 25 years old with no health problems except a few wore eye glasses. We

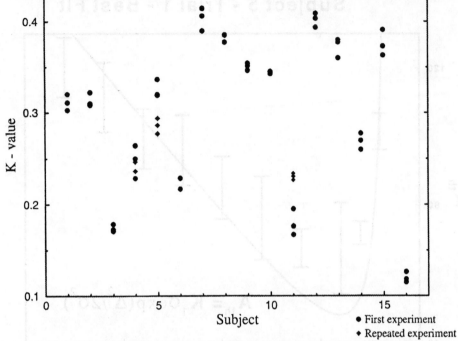

Fig. 11. Subject sensitivity, or K-value, versus subject number for all 16 subjects. The diamonds are the results of retesting the same subject after approximately one year.

believe that this test could be adapted to serve as a diagnostic of possible impairments of the human visual system. Tests of this hypothesis are currently in progress.

4. Discussion and Summary

We have discussed SR above as it is observed in single sensory neurons (Wiesenfeld and Moss, 1995; Bulsara, et al., 1991; 1996) in more complicated networks of neurons (the ganglia, which are, nevertheless, biologically rather simple) and finally as a tool for the quantifiable measurement of a performance task imposed on the human visual system. A previous computational work has also demonstrated SR at the retinal-ganglion cell level in a model of the human visual system (Stemmler, et al., 1995). These results herald the movement of SR into human medical science. Indeed, in a series of recent experiments, SR has been demonstrated in the human median nerve (Chiou-Tan, et al., 1996, 1997 and in press) and in human muscle spindles (Cordo, et al., 1996). It has come closest to a practical application in the use of noise to enhance the encoding performance of cochlear implants which may be usefully developed for the profoundly deaf (Morse and Evans, 1996).

Moreover, studies of SR in networks are important (Collins, *et al.*, 1995; Pantazelou, *et al.*, 1993; Gluckman, *et al.*, 1996) since the results can be used to interpret experimental data in higher level nervous systems. SR has also been generalized to excitable systems, which are often accurate descriptions of neurons (Moss, *et al.*, 1993; Collins *et al.*, 1995; Longtin, 1993; Wiesenfeld, *et al.*, 1994). Spatiotemporal SR has recently been introduced (Lindner, *et al.*, 1996; Hunt, *et al.*, 1996; Vilar and Rubi, 1997; Bulsara, *et al.*, 1994)). Some beautiful experiments on tactile SR in both humans (Collins, *et al.*, 1996a) and rat (Collins, *et al.*, 1996b) have recently been reported using aperiodic SR. Other psychophysical experiments have also been accomplished with human subjects (Chialvo and Apkarian, 1993; Collins, *et al.*, 1996a) and a bistable model of human perception of ambiguous figures which also demonstrated SR has been developed (Riani and Simonotto, 1994).

SR has also been studied in the context of neuron action potential timing sequences (Longtin, *et al.*, 1991, 1994) and precision (Pei, *et al.*, 1996). It is being studied in single ion channels (Petracchi, *et al.*, 1994) and arrays of ion channels (Bezrukov and Vodyanoy, 1995, 1997). Noise enhanced performance of subthreshold noisy oscillators in electrosensory organs are also a topic of current interest (Braun, *et al.*, 1994; Longtin and Hinzer, 1996).

The diversity of applications and wide variety of techniques indicated that SR is a fruitful and rewarding field of study for both theorists and experimentalists. Many open questions remain. These will be studied and many will be answered, in the fullness of time, by enthusiastic and outstanding researchers, both those cited here and others who will come after us.

Acknowledgements

SR research at the Center for Neurodynamics at the University of Missouri at St. Louis is supported by the U.S. Office of Naval Research, Physics Division.

References

Adey, W.R (1972) "Organization of brain tissue: is the brain a noisy processor?". *Intern. J. Neurosci.* **3**, 271-284.

Benzi, R. A. Sutera, and A. Vulpiani (1981) "The mechanism of stochastic resonance". *J. Phys. A* 14, L453-L457

Bezrukov, S.M. and I. Vodyanoy (1995) "Noise-induced enhancement of signal transduction across voltage-dependent ion channels". *Nature* 378, 362-364

Bezrukov, S.M. and I. Vodyanoy (1997) "Stochastic resonance in non-dynamical systems without response thresholds". *Nature* 385, 319-321

Braun, H.A., H. Wissing, K. Schäfer, and M.C. Hirsch (1994) "Oscillation and noise determine signal transduction in shark multimodal sensory cells". *Nature* 367:270-273.

Bulsara, A., E.W: Jacobs, T. Zhou, F. Moss and L. Kiss (1991) "Stochastic resonance in a single neuron model: Theory and analog simulation". *J Theor Biol.* 152: 531-555

Bulsara, A.R., S.B. Lowen, and C.D. Rees (1994) "Cooperative behavior in the periodically modulated Wiener process". *Phys. Rev. E.* 49: 4989-5000

Bulsara, A.R., T.C. Elston, C.R. Doering, S.B. Lowen, and K. Lindenberg (1996) "Cooperative behavior in periodically driven noisy integrate-fire models of neuronal dynamics". *Phys. Rev. E.* 53: 3958-3969

Chialvo, D.R. and A.V. Apkarian (1993) "Modulated noisy biological dynamics: three examples". *J. Stat. Phys.* 70:375-391

Chiou-Tan, F.Y., K. Magee, L. Robinson, M. Nelson, S. Tuel, T. Krouskop, and F. Moss (1996). Enhancement of subthreshold sensory nerve action potentials during muscle tension mediated noise. *Intern. J. Bifurc. And Chaos* 6: 1389-1396

Chiou-Tan, F., K. Magee, L.R. Robinson, M.R. Nelson, S.S. Tuel, A. Krouskop, and F. Moss (1997) "Augmented sensory nerve action potentials during distant muscle contraction". *Am. J. Phys. Med. Rehabil.*, 76: 14-18.

Chiou-Tan, F., T.Y. Chuang, T. Dinh, L.R. Robinson, S.S. Tuel, and F. Moss (1997) "Effect of nerve block on sural amplitude during remote muscle contraction". *Electromyogr. Clin. Neurophys.*, in press

Collins, J.J., C.C. Chow, and T.T. Imhoff (1995) "Aperiodic stochastic resonance in excitable systems". *Phys. Rev. E.* 52: R3321-R3324

Collins, J.J., T.T. Imhoff, and P. Grigg (1996a) "Noise-enhanced tactile sensation". *Nature* 383: 770

Collins, J.J., T.T. Imhoff, and P. Grigg (1996b) "Noise-enhanced information transmission in rat SA1 cutaneous mechanoreceptors via aperiodic stochastic resonance". *J. Neurophysiol.* 76: 642-645

Collins, J.J., C.C. Chow and T.T. Imhoff (1995) "Stochastic resonance without tuning". *Nature* 376: 236-238

Cordo, P., T. Inglis, S. Verschueren, J. Collins, D. Merfeld, S. Rosenblum, S. Buckley, and F. Moss (1996) "Noise in human muscle spindles". *Nature* 383: 769-770

Denk, W. and W. Webb (1992) "Forward and reverse transduction in the limit of sensitivity studied by correlating electrical and mechanical fluctuations in frog saccular hair cells". *Hear. Res.* 60: 89-102.

Douglass, J.K., L. Wilkens, E. Pantazelou, and F. Moss (1993) "Noise enhancement of information transfer in crayfish mechanoreceptors by stochastic resonance". *Nature* 365: 337-340

Gingl, Z., L.B. Kiss, and F. Moss (1995) "Non-dynamical stochastic resonance: theory and experiments with white and arbitrarily coloured noise". *Europhys. Lett.* 29: 191-196

Gluckman, B., T. Netoff, E. Neel. W. Ditto, M. Spano and S. Schiff (1996) "Stochastic resonance in a neuronal network from mammalian brain". *Phys. Rev. Lett.* 77: 4098-4101.

Hermann HT and R.E. Olsen (1967) "Dynamic statistics of crayfish caudal photoreceptors". *Biophysics J.* 7: 279-296.

Hibbs, A.D., A.L. Singsaas, E.W. Jacobs, A.R. Bulsara, J.J. Bekkedahl, and F. Moss (1995) "Stochastic resonance in a superconducting loop with a Josephson junction". *J. Appl. Phys.* 77: 2582-2590

Löcher, M., G. Johnson, and E.R. Hunt (1996) "Spatiotemporal stochastic resonance in a system of coupled diode resonators". *Phys. Rev. Lett.* 77: 4698-4602.

Huxley, T.H. (1880) *The Crayfish. An Introduction to the Study of Zoology.* D. Appleton, New York

Jung. P. (1993) "Periodically driven stochastic systems". *Phys. Rep.* 234: 175-295

Jung, P. (1997a) "Thermal waves, criticality, and self-organization in excitable media". *Phys. Rev. Lett.* 78: 1723-1726.

Jung, P. (1997b) "Kainate elicits noise-induced coherent fluorescence patterns in Astrocyte". *Syncytia*. Preprint.

Lanzara, E., R. Mantegna, B. Spagnolo, and R. Zangara (1997) "Experimental study of a nonlinear system in the presence of noise: The stochastic resonance". *Am. J. Phys.* 65:341-349.

Levin, J.E. and J.P. Miller (1996) "Broadband neural encoding in the cricket cercal sensory system enhanced by stochastic resonance". *Nature* 380: 165-168

Lindner, J., B. Meadows, W. Ditto, M. Inchiosa, and A. Bulsara (1996) "Scaling laws for spatiotemporal synchronization and array enhanced stochastic resonance". *Phys. Rev. E.* 53: 2081-2086

Longtin, A., A. Bulsara, and F. Moss (1991) "Time-interval sequences in bistable systems and the noise-induced transmission of information by sensory neurons". *Phys. Rev. Lett.* 67: 656-659

Longtin, A. and K. Hinzer (1996) "Encoding with bursting, subthreshold oscillations and noise in mammalian cold receptors". *Neural Comp.* 8: 217-257.

Longtin, A., A. Bulsara, D. Pierson, and F. Moss (1994) "Bistability and the dynamics of periodically forced sensory neurons". *Biol. Cybern.* 70: 569-578.

Longtin, A. (1993) "Stochastic resonance in neuron models". *J. Stat. Phys.* 70: 309-327

Morse, R. and E. Evans (1996) "Enhancement of vowel coding for cochlear implants by addition of noise". *Nature Medicine* 2: 928-932

Moss, F. (1994) "Stochastic resonance:from the ice ages to the monkey's ear". In *Contemporary Problems in Statistical Physics*, G. H. Weiss, ed., SIAM, Philadelphia, pp. 205-253.

Moss, F, D. Pierson, and D. O'Gorman (1994) "Stochastic resonance: Tutorial and update". *Intern. J. Bifurc. and Chaos*, 6: 1383-1397

Moss, F., J.K. Douglass, L. Wilkens, D. Pierson, and E. Pantazelou (1993) "Stochastic resonance in an electronic FitzHugh-Nagumo Model". *Ann. N.Y. Acad. Sci.* 706: 26-41

Moore, G., D. Perkel and J. Segundo (1966) "Statistical analysis and functional interpretation of neuronal spike data". *Ann. Rev. Physiol.* 28: 493-522.

Nicolis, C. (1982) "Long term climatic transitions and stochastic resonance". *Tellus*, 34: 1-9

Nicolis, C. (1993) "Long term climatic transitions and stochastic resonance". *J. Stat. Phys.* 70: 3-14.

Pantezelou, E., C. Dames, F. Moss, J. Douglass, and L. Wilkins (1995) "Temperature dependence and the role of internal noise in signal transduction efficiency of crayfish mechanoreceptors". *Intern. J. Bifurc. And Chaos* 5: 101-108

Pantezelou, E., F. Moss, and D. Chialvo (1993) "Noise sampled signal transmission in an array of Schmitt triggers". In *Noise in Physical Systems and 1/f Fluctuations*, AIP Press, New York pp. 549-552.

Pei, X., L. Wilkens and F. Moss (1996) "Light enhances hydrodynamic signaling in the multimodal caudal photoreceptor interneurons of the crayfish". *J. Neurophysiol.* 76: 3002-3011.

Pei, X., L. Wilkens and F. Moss (1996) "Noise-induced spike timing precision from aperiodic stimuli in an array of Hodgkin-Huxley-type neurons". *Phys. Rev. Lett.* 77: 4679-4682

Petracchi, D., M. Pellegrini, M. Pellegrino, M. Barbi, and F. Moss (1994) "Periodic forcing of a K+ channel at various temperatures". *Biophys. J.* 66: 1844-1852.

Riani, M. and E. Simonotto (1994) "Stochastic resonance in the perceptual interpretation of ambiguous figures: a neural network model". *Phys. Rev. Lett.* 72: 3120-3123.

Simonotto, E., M. Riani, C. Seife, M. Roberts, J. Twitty, and F. Moss (1997) "Visual perception of stochastic resonance". *Phys. Rev. Lett.* 78: 1186-1189

Stemmler, M., M. Usher, and E. Niebur (1995) "Lateral interactions in primary visual cortex: a model bridging physiology and psychophysics". *Science* 269: 1877-1880

Vilar, J.M.G. and J. M. Rubi (1997) "Spatiotemporal stochastic resonance in the Swift-Hohenberg equation". *Phys. Rev. Lett.* 78: 2886-2889.

Wiesenfeld, K. and F. Moss (1995) "Stochastic resonance and the benefits of noise: from the ice ages to crayfish and SQUIDs". *Nature* 373: 33-36

Wiesenfeld, K., D. Pierson, E. Pantazelou, C. Dames and F. Moss (1994) "Stochastic resonance on a circle". *Phys. Rev. Lett.* 72: 2125-2129

Wilkens L. (1988) "The crayfish caudal photoreceptor: Advances and questions after the first half century". *Comp Biochem Physiol* 91: 61-68

Wilkens, L and J. Douglass (1994) "A stimulus paradigm for analysis of near-field hydrodynamic sensitivity in crustaceans". *J. Exp. Biol.* 189: 263-272.

NONLINEARITY AND CHAOS IN NEURONAL AND CARDIAC DYNAMICS

TRANSMISSION OF A CHAOTIC SIGNAL BY A SYNAPTIC-LIKE COUPLING

SANTI CHILLEMI, MICHELE BARBI, ANGELO DI GARBO
Istituto di Biofisica del CNR, Via S. Lorenzo 26,56127-Pisa (Italy)
e-mail: chillemi@ib.pi.cnr.it

ABSTRACT

The sequences of interspike intervals generated from two integrate-and-fire neural models coupled in a synaptic-like way and in the presence of chaotic forcing are considered to investigate the process of information transmission from the first to the second oscillator, mainly looking at its dependence on the oscillators firing rates. The rationale is using the nonlinear predictability of interval sequences to monitor the information about the chaotic signal. The results show that information about the nonlinear signal is critically transmitted through the synaptic coupling and is maintained only under particular conditions.

1. Introduction

Neurons can be described as nonlinear, nonequilibrium systems subject to several inputs. Due to such intrinsic nonlinearity they are able to exhibit a large variety of behaviours ranging from deterministic to stochastic phase-locking up to a possible chaotic activity. This has led neuroscientists to consider neural time series as products of dynamical systems and thus to examine issues more general than those addressed by specific biological preparations. On the theoretical side, a number of techniques of time series analysis have been developed. The use of delay coordinates to reconstruct the state space of a dynamical system from the measurement of a scalar on the system itself (Takens, 1981), has proved a very powerful analytical tool. Moreover, it was shown that under certain genericity conditions a similar reconstruction holds when using sequences of interspike intervals (Sauer, 1994), as done in this report.

Experimentally, besides other types of complex dynamics, a clear nonlinear predictability of the interspike interval (ISI) sequence has been demonstrated in several preparations (Chang *et al.*, 1994; Schiff *et al.*, 1994; Hoffman *et al.*, 1995; Chillemi *et al.*, 1997). Nonlinear predictability is one of the fundamental properties - a dynamical invariant - of a deterministic system, like the correlation dimension or the Lyapunov exponents.

Now, since the transmission of information in the nervous system is based on synaptic coupling, the ability of a synaptic-like code to transmit nonlinear signals must be checked, both theoretically and experimentally. This paper tries to contribute to the theoretical investigation.

A simple model system is used herein to investigate these aspects: it consists of two Integrate-and-Fire (IF) models, with the first one being fed by a chaotic signal and the second one pulse-coupled to the first in inhibitory or excitatory way (at present, we are not interested to describe the synaptic coupling in a more realistic way). Once run the system and generated the two ISI sequences, the flow of information about the chaotic signal is followed along the brief chain by examining the short term nonlinear predictability of the same sequences (Schiff *et al.*, 1996).

The analysis method of nonlinear prediction (Farmer & Sidorowich, 1987; Sugihara & May, 1990) is able to detect the short-term predictability of ISI series even quite short (a few hundreds intervals). The underlying idea is that, if there is any level of determinism in a system, it should manifest itself as same predictability of its evolution (Longtin, 1993). Furthermore, to exclude that this predictability can be accounted for by a linear stochastic process, its statistical significance is assessed by using surrogate data (Theiler *et al.*, 1992).

The "experimental" protocols will be chosen in order to clarify how the transmission of information from the first to the second oscillator through the synapse depends on the relevant parameters, mainly the firing rates of the two oscillators.

2. Methods

2.1. Description of the system

The investigated system consists of three coupled elements: two Integrate-and-Fire oscillators, IF1 and IF2, where IF1 is fed by a chaotic signal obtained by integrating the Lorenz equations and influences IF2 through a synaptic coupling. The x-variable from the Lorenz equations

$$\frac{dx}{dt} = -\sigma(x-y),$$
$$\frac{dy}{dt} = -y + r x - x z,$$
$$\frac{dz}{dt} = -b z + x y,$$

with the standard parameter values $(\sigma, r, b) = (10, 28, 8/3)$ is used for supplying the chaotic signal. On the other hand, the integrate-and-fire neural model simply integrates the input signal up to a unity threshold whenever a spike is fired and the integral, or "potential", is reset to zero. The input signals to the two oscillators are given respectively by

$$v_1(t) = V_1 + A\,x\,(t),$$

where A is an amplitude parameter, and

$$v_2(t) = V_2 + I(t),$$

where $I(t)$ is a dicotomic signal, consisting of pulses of fixed amplitude (positive or negative), occurring at the moments of the spikes fired by the first oscillator. The product of this amplitude times the length of the integration step (see below) equals the simultaneous change of the IF2 potential and will be called *str*. It is worth noting that the free-run discharge rate r_{0i} of each oscillator numerically equals V_i (i = 1, 2).

The time evolution of both the Lorenz signal and the IF1 and IF2 potentials was calculated by simultaneously integrating the corresponding equations by using the second order Runge-Kutta and the Euler method, respectively. A step $\Delta t = 5 \times 10^{-4}$ was used. Fig. 1 shows the behaviour of the system in a typical case.

Thus, ISI sequences were generated for the nonlinear analysis. They contained from a minimum of 512 up to 4096 intervals and corresponded to time durations ever longer than 100 s.

2.2. Nonlinear Prediction and Surrogate Data Method

Given a physical system, measuring a single variable on it at equispaced intervals

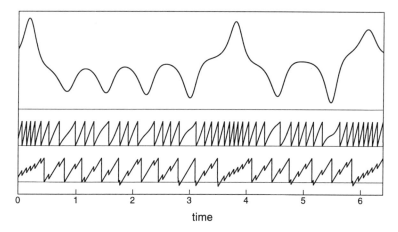

Figure 1. Typical time evolution of the model system described in the text. Upper: Lorenz signal; middle: first oscillator potential; lower: second oscillator potential.

yields a univariate (scalar) time series $X_1, X_2, \ldots, X_n$. A faithful reconstruction of the dynamics can be obtained by embedding the time series in a Euclidean space with sufficiently large dimension m (Takens, 1981; Sauer et al., 1991). This is usually achieved by forming vectors with the delayed values of the observable, namely the *delay vectors* $Xj = (X_j, X_{j-\tau}, \ldots, X_{j-(m-1)\tau})$ in R^m (the *lag time* τ is an integer).

Let us outline now the approach used for the analysis of the generated time series, i.e. the classical nonlinear prediction (Farmer & Sidorowich, 1987; Sugihara & May, 1990; Longtin, 1993), strengthened by the comparison with suitable surrogates (Theiler et al., 1992).

Let $X_1, X_2, \ldots, X_n$ be the series of ISIs and m the embedding dimension such that the *delay vectors* $Xj = (X_j, \ldots, X_{j-(m-1)\tau})$ in R^m allow the trajectory to be disentangled. To obtain the prediction $X_{i,H}$ at the time horizon H from *predictee* X_i, a given number (usually six) of the nearest-neighbours not sharing any coordinate with X_i are followed in their time evolution and the average of their translations after H time steps, with weights proportional to their reciprocal distances from X_i, is computed. This gives $X_{i,H}$. Then as many predictees as possible are selected in the reconstructed space (Sugihara & May, 1990) and the corresponding predictions are calculated. The *out-of-sample predictability* of the time series $\varepsilon(H)$ is quantified by the adimensional ratio between the average prediction error and the series standard deviation σ. Note that σ can be regarded as the average prediction error obtained by assuming the mean of the time series as the same prediction for all times (Kaplan & Glass, 1995). Explicitly:

$$\varepsilon(H) = \frac{1}{\sigma} \sqrt{\frac{\sum_i (X_{i,H} - X_{i+H})^2}{N}} \qquad (1)$$

where N is the number of predictees. This statistic will be referred to in the following as the *normalised prediction error* (NPE). Obviously, values of $\varepsilon < 1$ clearly imply some predictability, linear and/or nonlinear. However, even ε-values slightly larger than unity should be assumed to mean predictability, when the NPEs of the original and the surrogate series rise, for high H values, to a common level higher than unity (see the Appendix).

In the analysis reported here, the value at which the autocorrelation function decays below $1/e$ was generally adopted for the lag time τ. Moreover, to choose the value of the embedding dimension m, we considered that the attractor dimension of the ISI sequence of the simple Lorenz-driven IF model should be similar to that of the Lorenz attractor, of about 2 (Sauer, 1994; Racicot & Longtin, 1997). This leds us to adopt $m = 5$ for the ISI sequences of both oscillators. Moreover, a check of the m-dependency of the one-step-ahead NPE showed the choice done to be good.

Now, the mere observation of a short term predictability in the data is not sufficient to assess the low-dimensionality of the underlying system; coloured noise containing significant autocorrelation generates the same behaviour. As anticipated in the introduction, to exclude that the data predictability can be accounted for by a linear stochastic process, statistical controls based on surrogate data must be carried out. The *Fourier shuffled* surrogates used here have the same mean, standard deviation, and amplitude histogram as the original sequence, and a slightly changed autocorrelation (Theiler *et al.*, 1992; Chang *et al.*, 1994).

Once the surrogate sequences are created, their prediction error is computed and compared with that of the original series: a statistically meaningful difference invalidates the *null hypothesis* that the data predictability is accounted for by linear correlation in time. At the same time, the possible difference estimates the original series *nonlinear predictability*, thereby separated from the *linear predictability* which is exhibited by the surrogate series.

3. Results

The features of information transmission from the first to the second oscillator through the synaptic coupling will be investigated by inspecting the predictability curves (ε against H) calculated on the sequences of ISIs of the second oscillator and their Fourier shuffled surrogates. Accordingly, the curves obtained from sequences generated in different operation conditions will be shown in the following. However, it is worth showing first how the first oscillator discharge is able to encode the chaotic dynamics of the Lorenz signal. The predictability curves plotted in Fig. 2, a,b,c, refer to the rates $r_{01} = 2$, 4 and 8 sp/s respectively. They show the dependence

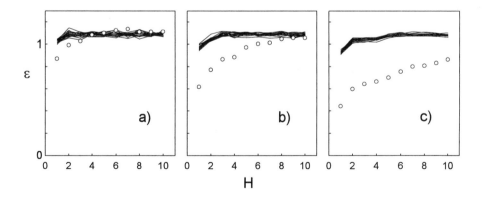

Figure 2. Plot of ε against prediction time for the original ISI sequence of IF1 (open circles) and the surrogate data (continuous lines): a) $A=0.1$, $r_{01}=2$ sp/s; b) $A=0.2$, $r_{01}=4$ sp/s; c) $A=0.4$, $r_{01}=8$ sp/s.

on prediction time of the original series NPE and the NPEs of the nineteen surrogate ones. With this particular choice of the surrogate number, "experimental" prediction errors smaller than any of the surrogate ones have significance at the 0.05 probability level (Chang et al., 1995; Schiff et al., 1996). In the plots of Fig. 2, the short-term nonlinear predictability strongly increases with the firing rate, for this reason the curves at higher rates are not presented. This is well in keeping with other's results (Sauer, 1994; Racicot & Longtin, 1997).

Let us come now to the second oscillator and consider firstly the case of inhibitory coupling ($str < 0$). Fig. 3 shows the predictability changes of its discharge as the first oscillator rate is increased (plots a and b in Fig. 3 correspond to the same first oscillator discharge as plots b and c in Fig. 2) and simultaneously the strength str of the synaptic quantum is proportionally decreased, in order to keep approximately constant the average firing rate of the second oscillator. This protocol eliminates the effect on the information transmission of the lengthening of the second oscillator ISI interval that would occur if the quantum strength was kept constant and allows us to directly follow the increase in resolution of the driving chaotic signal provided by the synaptic coupling. In the following, the firing rate of the uncoupled second oscillator will be referred to as r_{02}, while its real rate in the presence of coupling will be r_{02}'. So for Fig. 3 the nominal rate of $r_{02} = 8$ sp/s is reduced to $r_{02}' = 6.4$ sp/s by the inhibitory coupling.

Fig. 4 shows the nonlinear predictability curves of the ISI sequence of the second oscillator as its own discharge rate changes, while the rate of the first oscillator is kept constant at $r_{01} = 32$ sp/s. Now the strength of the synaptic quantum is changed inversely with respect to r_{02}, in order to keep constant the inhibitory influence of IF1 (percent reduction of the IF2 rate). While the nonlinear signal is well transmitted for the intermediate rate ($r_{02} = 8$ sp/s), for the two extreme ones it is strongly reduced.

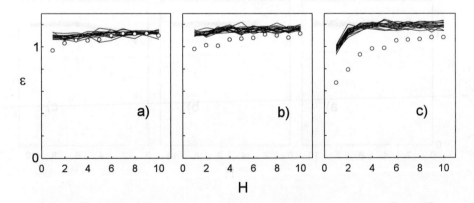

Figure 3. Plot of ε against prediction time for the original ISI sequence of IF2 (open circles) and the surrogate data (continuous lines) for $r_{02}=8$ sp/s and: a) $A=0.2$, $r_{01}=4$ sp/s, $str=-0.4$; b) $A=0.4$, $r_{01}=8$ sp/s, $str=-0.2$; c) $A=0.8$, $r_{01}=16$ sp/s, $str=-0.1$.

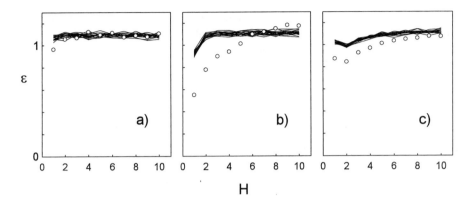

Figure 4. Plot of ε against prediction time for the original ISI sequence of IF2 (open circles) and the surrogate data (continuous lines) for $A=0.8$, $r_{01}=32$ sp/s and: a) $r_{02}=2$ sp/s, $str=-0.025$; b) $r_{02}=8$ sp/s, $str=-0.1$; c) $r_{02}=32$ sp/s, $str=-0.4$.

Two different reasons can account for this reduction. At the lowest nominal rate ($r_{02} = 2$ sp/s), the actual average ISI of 0.82 s approaches the Lorenz signal decorrelation time, so impairing the transmission of the dynamic information; instead, at the highest rate ($r_{02} = 32$ sp/s), the second oscillator ISIs become so short that the synaptic quanta falling within any of them are very few whereby little different in number. In conclusion, ISIs loose resolution (the ISI sequence becomes *course grained*) and that clearly reduces the quality of the transmitted signal.

It is worth noting that the simple halving of the input signal, i.e. the reduction of the r_{01} level from 32 sp/s to 16 sp/s and of the amplitude A of the modulating signal

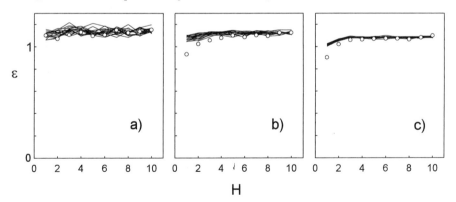

Figure 5. Plot of ε against prediction time for the original ISI sequence (open circles) of IF2 and the surrogate data (continuous lines) for the same parameter values as in Fig. 4 except for $r_{01}=16$ sp/s and $A=0.4$.

from 0.8 to 0.4, causes the above effects almost disappear, even maintaining some short range nonlinear predictability at the two highest frequencies (Fig. 5). Also this change can be accounted for by means of the loss of resolution of the second oscillator ISIs caused by the decrease in modulation of the first oscillator ISI sequence.

To further demonstrate the importance of the resolution in the interspike intervals, Fig. 6 shows the predictability curves of the second oscillator discharge as **only** the Lorenz signal amplitude is changed. The nonlinear predictability clearly increases with that amplitude, what can be ascribed to the increase of the range of values swept by the second oscillator ISIs.

Finally, by changing the type of coupling from inhibitory to excitatory, one should expect the general behaviour not to show qualitative changes. However, by running a simulation using all the same parameter values as in Fig. 5, but reversing the sign of *str*, a quite higher nonlinear predictability comes out, as shown in Fig. 7. This enhancement is particularly evident in Fig. 7c and it can be accounted for by considering the recovery of amplitude resolution due to the immediate triggering of a spike of the second oscillator every time a quantum falls in the last part of the ISI. Also, it should be taken into account that the shortening of intervals produced by the excitatory quanta acts as a stabilising process of negative feedback. But further investigations are needed to definitely validate these explanations.

4. Discussion

Synaptic coupling can be seen as the elementary process underlying talking between neurons; so understanding synapses is a necessary - but not sufficient - step towards understanding neural processing. In this report the conditions under which

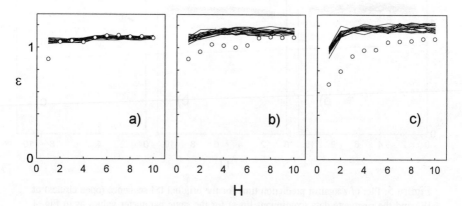

Figure 6. Dependence of the information transmission on the modulation amplitude. Parameter values: r_{01}=16 sp/s, str=-0.1, r_{02}=8 sp/s; a) A=0.2; b) A=0.4; c) A =0.8.

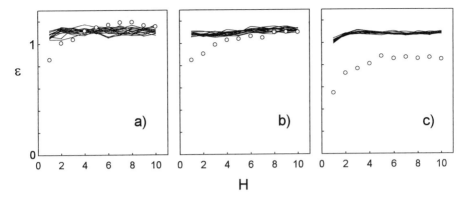

Figure 7. Plot of ε against prediction time for the original ISI sequence (open circles) of IF2 and the surrogate data (continuous lines) for the same parameter values as in Fig. 5 except that the coupling is now excitatory.

a simple model of synapse is able to transmit the information carried from a chaotic signal, have been investigated.

In fact, one open question is how much nonlinear predictability of the input signal to a neuron can be recovered in principle from the ISI sequence.

Now, it is well known that, for the simple IF neural model, the dynamical invariants of a chaotic driving signal (included the nonlinear forecastability), are better maintained at higher firing rates. In fact, when the firing rate is too low, the information about the fine structure of the input signal cannot be encoded in the discharge (Sauer, 1994; Racicot & Longtin, 1997). The predictability curves computed on the first oscillator discharge and plotted in Fig. 2 confirm this point. Instead, the behaviour of the second oscillator, synaptically inhibited from the first one, appears more complicated: here the information about the chaotic signal is maximum for a window of firing rates. Such finding implies that the neural discharge allows the dynamical invariants of the driving signal to be recovered only for suitable firing rates (with respect to the rate of the synaptic bombardment).

The optimisation of the information transmission through the inhibitory synapse at the "central" firing rates can be accounted for qualitatively. The waste of information at the lowest firing frequencies is due to the same reason as for the first oscillator. Instead, the high rate cutoff is caused by the loss of resolution of the output ISIs due to the very few quanta falling in an interspike interval. Moreover, as Fig. 5 shows, a reduction of ISI resolution can occur also for low values of r_{02} or better of ratio r_{02}/r_{01}, as long as the amplitude of the modulating signal is small enough.

Changing the type of coupling from inhibitory to excitatory, brings about qualitative changes of the system's response. This is probably due to the ability of excitatory quanta to instantaneously trigger a spike in the "follower" neuron. On the other hand, the behaviour of networks containing excitatory neurons is still an open problem (Budelli *et al.*, 1997).

In conclusion, the scheme of synaptic coupling adopted here, although oversimplified, can illustrate many important features of the information transmission about the nonlinear driving signal.

Acknowledgments

This research was supported by the National Research Council of Italy within the project "Spatio-Temporal Chaos in Biological Excitable Systems".

Appendix

Here, we will show that, for a time series lacking any predictability, or also for a generic series at rather large H values, ε rises to a value L larger than unity, so the right evaluation of the prediction curves must refer to the interval $(0,L)$ as to the ε domain.

Let us consider the simplest version of the prediction algorithm when the prediction at time H in the future is obtained by the corresponding evolution of only one nearest-neighbour (a situation that approximates the case of several nearest-neighbours, as long as their evolutions are averaged with weights proportional to the reciprocals of the respective distances from the predictee). Suppose also that the series $x_1, x_2, \ldots, x_n$, is a *white sequence* with IID x-variables uniformly distributed in the interval $(0,1)$. Then it is easy to show that $\sigma = 1/\sqrt{12}$ and the *rms* of the prediction error is $\sqrt{2}$ times larger, or $\varepsilon = \sqrt{2}$ (the predictions are completely random and with the same marginal distribution as x).

On the other hand, if k equally weighted nearest-neighbours are followed to make the prediction, the value $f(k) = \sqrt{(k+1)/k}$ is obtained for ε.

Lastly, in the already quoted (mentioned) case of weights decreasing with the distance from the predictee, ε will be intermediate between $f(k)$ and $\sqrt{2}$.

Coming now to a generic x-distribution (e.g. gaussian), things will be a little different but still the previous results approximately hold for high time horizons. In conclusion, when both the original and the surrogate series prediction errors for high H values rise to $L > 1$, it is this asymptotic value that one must refer to as corresponding to a vanishing predictability.

References

Budelli R., E. Catsigeras, A. Rovella & L. Gomez (1997) "Dynamical behavior of pacemaker neurons networks", *Nonlinear Analysis, Theory, Methods & Applications* 30: 1633-1638.

Chang, T., T. Sauer, S.J. Schiff (1995) "Tests for nonlinearity in short stationary time series", *Chaos* 5: 118-126.

Chang, T., S.J. Schiff, T. Sauer, J.P. Gossard & R.E. Burke (1994) "Stochastic Versus Deterministic Variability in Simple Neuronal Circuits: I. Monosynaptic Spinal Cord Reflexes", *Biophys. J.* 67: 671-683.

Chillemi, S., M. Barbi & A. Di Garbo (1997) "Dynamics of the neural discharge in snail neurons", Biosystems 40: 21-28.

Farmer, J.D. & J.J. Sidorowich (1987) "Predicting chaotic series", *Phys. Rev. Lett.* 59: 845-848.

Hoffman R E., W. Shi & B.S. Bunney (1995) "Nonlinear Sequence-Dependent Structure of Nigral Dopamine Neuron Interspike Interval Firing Patterns", *Biophys. J.* 69: 128-137.

Kaplan, D.T. & L. Glass (1995) "Understanding nonlinear dynamics", Springer Verlag.

Longtin, A. (1993) "Nonlinear forecasting of spike trains from sensory neurons", *Int. J. Bif. and Chaos* 3: 651-661.

Racicot, D.M. & A. Longtin (1997). "Interspike interval attractors from chaotically driven neuron models", *Physica D* 1595: 1-21.

Sauer, T. (1994) "Reconstruction of dynamical systems from interspike intervals", *Phys. Rev. Lett.* 72: 3811-3814.

Schiff, S.J., K. Jerger, T. Chang, T. Sauer & P.G. Aitken (1994) "Stochastic versus deterministic variability in simple neuronal circuits: II Hippocampal slice", *Biophys. J.* 67: 684-691.

Schiff, S.J., P. So, T. Chang, R.E. Burke & T. Sauer (1996) "Detecting dynamical interdependence and generalized synchrony through mutual prediction in a neural ensemble", Phys. Rev. E 54: 6708-6724.

Sugihara, G. & R. May (1990) "Nonlinear forecasting as a way of distinguishing chaos from measurement error in a data series", *Nature* 344: 734-741.

Takens, F. (1981) "Dynamical systems and turbulence", in: *Lecture Notes in Math.* D.A. Rand & L.S. Joung eds. Vol. 898: Springer-Verlag.

Theiler, J., B. Galdrikian, A. Longtin, A. Eubank & J.D. Farmer (1992) "Testing for nonlinearity in time series: the method of surrogate data", *Physica* D 58: 77-94.

SYSTEM RECONSTRUCTION AND ANALYSIS USING INTERSPIKE INTERVALS

ROLANDO CASTRO and TIM SAUER
*Institute of Computational Sciences and Informatics
and Department of Mathematical Sciences
George Mason University
Fairfax, VA 22030, USA
tsauer@gmu.edu*

ABSTRACT

Statistical measures of determinism for spikes train are described and analyzed. The analysis is focused on two dynamical invariants: nonlinear forecastability and correlation dimension. Two simple, generic, biologically motivated methods for generating spikes from the system are surveyed: an integrate-and-fire model (IF) and a threshold-crossing method (TC). The effect on the statistical measures of threshold level, noise and limited amount of available data are investigated.

1. Introduction

Substantial progress has been made on chaotic time series analysis during the last twenty years. Identification and reconstruction of the chaotic system are specific goals of the researchers, which underlie the applications of system prediction and control. Using delay coordinates (Packard et al., 1980; Takens, 1981) to reconstruct the state space of the dynamical system has been a successful idea. According to (Sauer et al., 1991), a delay coordinate reconstruction of a compact attractor from a time series is topologically equivalent to the attractor for a probability-one choice of measurement functions, as long as the embedding dimension is greater than twice the box-counting dimension of the attractor. This topological equivalence is the basis for a great deal of the statistical methodology for treating chaotic time series.

There is an analogous problem of relating interspikes intervals (ISI) to the dynamical system from which they are measured. This problem arises in many physiological systems when direct measurements of the state variables, to obtain an evenly-sampled time series, are not available or possible. Some researchers have attempted to adapt methods of chaotic time series analysis to single neuron or neuron population spike trains, others to the description of more global electrophysiological measurements such as the EEG.

Many mathematical models have been proposed for the dynamical structure of spiking phenomena in physiological systems (Tuckwell, 1988). In this report we survey two of these models and discuss the possibility of recovering the dynamical invariants of the system from the spike train. In particular we focus on the correlation dimension and nonlinear predictability of the ISI's, and show how they can be used for control purposes. The goal of this article is to study how evidence of deterministic nonlinear dynamics can be carried by spike trains.

In section 2 we describe the surveyed models: integrate and fire (IF) and threshold crossings (TC). In section 3 we focus our attention on the calculation of the correlation dimension of the attractor using the ISIs for both methods. In section 4 the predictability of the ISIs is investigated, and methods of subthreshold and superthreshold control from spike trains are surveyed in Section 5. Section 6 discusses the limitations of this study and directions for future work.

2. Simple models for spike generation

In this section we describe the methodology used for producing spike timing information from a dynamical system. Two models were chosen for their simplicity and genericity.

2.1. Integrate and fire model

In the integrate and fire model (IF) studied in (Sauer, 1994), a signal produced by a time-varying function of the variables of a low-dimensional dynamical system is integrated with respect to time until it reaches a preset threshold. At that time a spike is generated and the integration is restarted. Assume that the trajectories of the dynamical system are asymptotic to a compact attractor. Let $S(t)$ represent the signal and Θ the firing threshold. After fixing a starting time T_0, a series of "firing times" $T_1 < T_2 < T_3 < \ldots$ can be recursively defined by the equation

$$\int_{T_i}^{T_{i+1}} S(t)dt = \Theta. \tag{1}$$

From the firing times T_i, the interspike intervals (ISI) can be defined as $t_i = T_i - T_{i-1}$. Figure 1 shows a trace of the x coordinate of the Lorenz attractor (Lorenz, 1963), governed by the equations $\dot{x} = \alpha(y-x) \dot{y} = \rho x - y - xz \dot{z} = -\beta z + xz$, where the parameters are set to the standard values $\alpha = 10, \rho = 28, \beta = 8/3$, together with spiking times generated by (1) where $S(t) = x(t) + 25$ and $\Theta = 10$. This choice for $S(t)$ is representative. A wide range of positive functions $S(t)$ yielded results similar to those about to be described. The threshold Θ, on the other hand, turns out to be a critical parameter, and is discussed below in more detail.

Under certain genericity conditions on the underlying dynamics, signal and threshold, the series $\{t_j\}$ of ISI's can be used to reconstruct the attractor. That is, there is a one-to-one correspondence between the m-tuples of ISI's and attractor states, which associates each vector $(t_i, t_{i-1}, \cdots, t_{i-m+1})$ of ISI's with the corresponding point $x(t_i)$ on the attractor. In analogy with Takens theorem (Takens,

1981) and its generalization (Sauer et al., 1991), the condition $m > 2D_0$ is sufficient, where D_0 is the box counting dimension of the attractor. As with Takens'

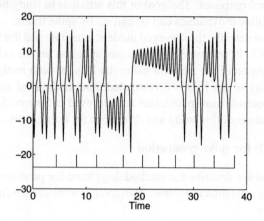

Figure 1: The upper trace $x(t)$ is the x coordinate of the Lorenz attractor graphed as a function of time. The lower trace shows the times at which spikes are generated using $S(t) = x(t) + 25$ and $\Theta = 10$.

theorem, smaller m may be sufficient in particular cases. See (Sauer, 1996) for a discussion of the proof of this fact for integrate-and-fire interspike interval data.

2.2. Threshold crossings

In the threshold crossings method (TC), the time intervals between successive, upward-going level crossings of the time series through a threshold Θ are recorded. Figure 2 shows the xy projection of the (Rössler, 1976) attractor governed by the equations $\dot{x} = -(y+z), \dot{y} = x + ay, \dot{z} = b + (x-c)z$, where the parameters are set to the standard values $a = 0.36, b = 0.4, c = 4.5$, together with the lines $x = 2$ and $x = -3$. Note that the firing occurs with the great majority of the oscillations of the attractor when $\Theta = 2$. This is not the case when $\Theta = -3$ where many of the smaller oscillations are missed. For $\Theta = -3$ much of the dynamical behaviour captured by the time series will be missed by the TC interspike intervals. We will see the effect of this loss of information in correlation dimension calculations below, and in a more precise way later in the context of forecastability.

3. Correlation Dimension

This section surveys the possibility of recovering the correlation dimension from a spike train. The results shown here are taken from (Castro & Sauer, 1997).

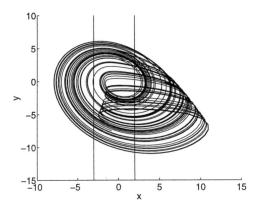

Figure 2: An xy projection of the Rössler attractor. The lines $x = 2$ and $x = -3$ represent possible firing thresholds Θ.

The correlation dimension D_2 of an invariant measure of a dynamical system is given by

$$D_2 = \lim_{r \to 0} \frac{\log C(r)}{\log r}, \tag{2}$$

where the correlation integral $C(r)$ is defined to be the probability that a pair of points chosen at random is separated by a distance less than r. Let $\{x_1, x_2, \ldots, x_N\}$ be the first N points of a trajectory which traces out the invariant measure. The correlation integral can be approximated by

$$C(N, r) = \#\{ \text{ pairs } x_i, x_j : |\ x_i - x_j\ | < r \ \}/(N(N-1)/2) \tag{3}$$

where $|\cdot|$ denotes the distance norm. As N goes to ∞, $C(N, r)$ goes to $C(r)$.

Given a sequence of ISIs we can construct an m-dimensional space using delay coordinates and compute $C(m, N, r)$ according to Eq. (3) for a given m. (Here a parameter m is added to indicate the dimensionality of the reconstruction space.) If the estimated values $\bar{D}_2^{(m)}$, plotted as a function of m, appears to reach a plateau for a range of large enough m, then the plateau dimension value is taken to be an estimate for D_2 (Ding et al., 1993)

In Fig. 2 we plot $\log C(m, N, r)$ versus $\log r$ for $m = 3$ to $m = 8$ and $N = 64000$. The ISIs were generated using the x coordinate of the Lorenz attractor, the signal $S(t) = x(t) + 25$ and $\Theta = 10$. An optimized box-assisted algorithm for correlation dimension (Grassberger, 1990) was used to compute $C(m, N, r)$.

To estimate D_2, (Grassberger and Procaccia, 1983) suggest that a log-log plot of $C(N, r)$ versus r be constructed and that the estimation of the dimension, denoted by $\bar{D}_2$, be read off as the slope of the curve over some range where the graph

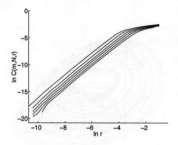

Figure 3: Log-log plot of the correlation integrals of the Lorenz ISI (signal $S(t) = x(t) + 25$, $\Theta = 10$) for m = 3 to m = 8. The top line is for m = 3, the bottom line m = 8.

show linear dependence. A variant of this method was suggested by (Takens, 1985). The Takens estimator requires the choice of a single free parameter R_0 the upper cutoff distance. All pairwise distances larger than R_0 are discarded, and all distances r which are less than R_0 are averaged according to

$$d = \frac{-1}{<\log(r/R_0)>}. \tag{4}$$

This d is an estimate for D_2. Using the Takens estimator of dimension makes our conclusions less sensitive to choice of scaling region in the determination of dimension, in particular for cases that are less clear cut than Fig. 3. We set the parameter R_0 equal to one-half of one percent of the diameter of the reconstructed attractor.

We illustrate the ability to recover correlation dimension through IF interspike intervals. Using a spike train generated from the IF hypothesis (1) with $S(t) = x(t) + 25$ from (2), we embed m-tuples of interspike intervals as m-dimensional vectors and calculate the Takens estimator of dimension using (6). Figure 4 shows the results of the calculations.

To investigate the effect of the size of the threshold in the IF model we compute the correlation dimension of ISI series generated with (1) using different thresholds. As the threshold increases, the length of time spanned by an ISI vector grows, and decorrelation due to sensitive dependence can damage the reconstruction.

In Fig. 4 we plot $\bar{D}_2^{(m)}$ versus m for $\Theta = 10$ and 50. The signal and trajectory used are the same as in Fig. 3. As shown in the Fig. 4, with $\Theta = 10$ (diamonds), the plot reaches a plateau close to the Lorenz correlation dimension 2.06. For $\Theta = 50$ we were not able to recover the Lorenz attractor dimension. Increasing the threshold by a factor of 5 results in an increase in the mean interspike interval of roughly 5. Referring to Fig. 1, where $\Theta = 10$, one concludes that the system moves through many oscillations during each ISI when $\Theta = 50$. Since the Lorenz attractor has a Lyapunov exponent of approximately 0.9, information about the original system state is lost after only one or two spike intervals. At the precision with which the ISIs were measured in this study, useful information about the attractor is not reproduced in the reconstruction. This problem for ISIs is analogous to the effect of large time delays when reconstructing an attractor from time series.

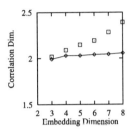

Figure 4: Dimension estimates of integrate-and-fire Lorenz system measured from interspike intervals, as a function of embedding dimension m. Takens estimator with $R_0 = 0.5$ percent of diameter is used. The diamond symbols marks dimension estimator for threshold $\Theta = 10$; squares mark threshold $\Theta = 50$.

Similar results are seen from IF interspike interval series generated from the Rössler attractor. For low thresholds the dimension can be derived correctly by embedding interspike intervals; for larger information is lost due to exponential divergence of trajectories.

To investigate the effect of the threshold in the TC model we compute the correlation dimension of interval series generated by upcrossings of the x-variable of the Rössler equations. As can be seen from Fig. 2 there are a range of threshold for which the smaller oscillations are missed. As in the IF model for large threshold, decorrelation effect due to sensitive dependence can limit the faithfulness of the reconstruction.

The information captured by the TC method differs from that of the IF method. In particular, the ISIs generated with the TC method can not be used to reconstruct the underlying attractor in the sense of (Sauer, 1994). Instead, these ISIs measure the times between piercings of a Poincaré surface of section, which suggest the conjecture that vectors consisting of m successive interspike intervals created in this way will generically comprise a set of dimension exactly one less than the attractor dimension.

In Fig. 5(a) we plot the dimension estimates, using the Takens estimator on m-tuples of ISIs as above, for upcrossings of the threshold $\Theta = 2$. Note that, as stated earlier, for this threshold the firing occurs with the great majority of the oscillations of the underlying attractor. The dimension estimates are close to 1; for the diamonds shown in Fig. 5(b), representing the dimension calculation using 64000 interspike intervals, the dimension estimate is 1.01, essentially equal to one less than the correlation dimension of the Rössler attractor as we conjectured. When fewer than 64000 intervals are used, the dimension estimate degrades in accuracy. The squares and triangles in Fig. 5 represent the dimension estimates for 16000 and 4000 intervals, respectively.

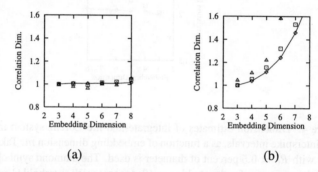

Figure 5: Dimension estimates of Rössler system measured from threshold-crossing interspike intervals, using Takens estimator. The diamond, square, and triangle symbols correspond to ISI sequence lengths 64000, 16000, and 4000, respectively. Threshold in the x variable is (a) $\Theta = 2$ (b) $\Theta = -3$.

The alternate threshold $\Theta = -3$ in Fig. 2 leads to a quite different result. As can be seen from Fig. 2, many of the smaller oscillations will now be missed by the threshold-crossing method. Although our conjecture is that in theory the correct dimension can be found from the $\Theta = -3$ upcrossings, in practice the data requirements are difficult to satisfy. The likely cause of the difference between thresholds $\Theta = 2$ and -3 is that the latter produces fewer spikes per unit time, because small oscillations are missed. For a fixed embedding dimension, the length of dynamical time spanned by a reconstructed ISI vector becomes comparable with the decorrelation time of the chaotic system, which has the effect of degrading the attractor reconstruction. Figure 5 shows the resulting inaccuracy in the dimension estimate. This is perhaps consistent with the findings of (Preissl et al., 1990), who reported difficulty in obtaining the correct correlation dimension of the Lorenz attractor from threshold-crossings, possibly due to this decorrelation effect. In that study, 5000 ISIs were used in the dimension calculation.

4. Nearest Neighbor Prediction

In (Sauer, 1994) the possibility of predicting future interspike intervals from the IF model was explored. A simple version of a nearest neighbor prediction algorithm was used. Given an ISI vector $I_0 = (t_{i_0}, \cdots, t_{i_0-m+1})$ from the time series $\{t_i : 1 \leq i \leq n\}$, k nearest neighbors are chosen. The k nearest neighbors are translated by the horizon H, and their average

$$p_{i_0} = \frac{1}{k} \sum_{j=1}^{k} t_{j+H},$$

is used to approximate the future interval t_{i_0+H}. Neighbors which are close in time are not considered for inclusion, so that the prediction considered here is essentially out-of-sample prediction. This prediction can be viewed as a *zero-order* prediction of the translation of t_0 by H time units. The difference $p_{i_0} - t_{i_0+H}$ is the H-step prediction error at step i_0. We could instead use the series mean M to predict at each step; this H-step prediction error is $M - t_{i_0+H}$. The ratio of the root mean square errors of the two possiblities gives the normalized prediction error

$$\text{NPE} = \frac{< (p_{i_0} - t_{i_0+H})^2 >^{1/2}}{< (M - t_{i_0+H})^2 >^{1/2}}.$$

If NPE = 0 the predictions are perfect; NPE = 1 indicates that the performance is not better than the mean constant predictor M.

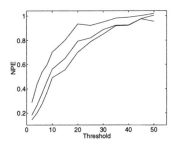

Figure 6: Normalized one step ahead prediction error for the IF model using the Lorenz equations for different amounts of additive noise added to the time series used to create the ISIs. The bottom line no noise is added. The top line is for 25 percent noise and the middle line is for 10 percent noise.

The results of (Sauer, 1994) showed that predictability beyond linear autocorrelation can be easily detected in IF interspike intervals created from single chaotic attractors. Therefore, prediction can be used as an indicator of nonlinear determinism for spike trains in the same manner that it is often used for measured time series.

In this work, we add in-band noise to the time series $\{t_i\}$ used to create the spikes timings. The in-band noise is a new time series $\{n_i\}$ with similar statistical properties as the original but which is a realization of a stochastic process, called a *surrogate time series* (Theiler et al., 1992). The new noisy time series is $u_i = t_i + n_i$, and the proportion of observational noise added is $\sigma_{n_i}/\sigma_{t_i}$, the ratio of the standard deviations of the noisy and clean series.

The result of applying the nearest neighbor prediction algorithm to integrate-and-fire ISI's from the Lorenz attractor is shown in Fig. 6. The signal $S(t) = x(t) + 25$ and trajectory used are the same as in Fig. 1. We fixed the embedding dimension $m = 3$, and predicted one step ahead ($H = 1$). Interval spike intervals series of length 1024 were generated with varying thresholds. The NPE is plotted as a function of the threshold for different amounts of observational noise $(0, 10, 25)$. Note that as the threshold increases the predictions degrade. Again this is the effect of the decorrelation mentioned in the previous section.

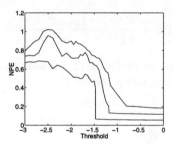

Figure 7: Normalized one step ahead prediction error for the TC model using the Rössler equations. The bottom line represents prediction error where no noise was added to the threshold-crossings process. The top line is for 10 percent noise and the middle line is for 5 percent noise.

The result of applying the nearest neighbor prediction algorithm to TC intervals from the Rössler attractor is shown in Fig. 7. As in the IF interval prediction, we fixed the embedding dimension $m = 3$, and predicted one step ahead ($H = 1$). Threshold crossing ISIs of length 1024 were generated with varying thresholds. The NPE is plotted as a function of the threshold for different amounts of observational noise $(0, 5, 10)$.

As shown in Fig. 2 there is a range of thresholds, $[-3, -1.5]$ approximately, that misses the small oscillations. It can be seen in Fig. 7, that for these range of thresholds the prediction is poor for the clean series (bottom line) and degrades when noise is added (top two lines). This is the effect of the decorrelation mentioned in the previous section. For the range of thresholds that does not miss the small oscillations, $[-1.5, 2]$ the NPE is small, the prediction is better. Note in particular the steep drop in prediction error as the upcrossing threshold moves above aproximately -1.5, in the noiseless case. The steepness of this drop is attenuated when noise is added to the process.

5. Control from interspike intervals

Stabilization of unstable periodic orbits in chaotic systems using small controls was described by (Ott et al. 1990), and is commonly referred to as OGY control. Several experimental examples of the success of this type of stabilization are documented in (Ott et al., 1994). The fundamental idea is to determine some of the low period unstable orbits embedded in the attractor, and then use small feedback to control the system. The OGY method is model-free, in that it relies on time series information to reconstruct the linear dynamics in the vicinity of the unstable orbit.

In this section we would like to prove the viability of this concept when the dynamical information is carried by interspike intervals instead of a time series. We will demonstrate two different types of control protocols. The first will be the analogue of OGY control for spike trains. Using small changes in a system parameter, an unstable orbit will be stabilized. This type of control is called subthreshold control, since the control signal is not large enough on their own to make the system fire. For this demonstration we will use spike intervals produced by an observable from the Lorenz system, and stabilize the figure-8 trajectory.

The second protocol will use superthreshold pulses to stabilize an unstable equilibrium. This technique, often called on-demand pacing, will be shown to work for deterministic as well as random integrate-and-fire systems. Therefore, although the success of subthreshold control can be viewed as evidence of deterministic dynamical behavior in an unknown system, the same cannot be said for superthreshold control.

We begin with subthreshold control. The standard version of the Lorenz attractor exhibits trajectories cycling erratically around two unstable equilibrium points. The number of times a typical trajectory cycles around one point before moving to the other is known to be unpredictable on moderate to long time scales. We will choose to stabilize the unstable periodic trajectory which cycles exactly once around one equilbrium, then exactly once around the other, before returning to the first and repeating the process. This "figure-eight" trajectory of the Lorenz attractor is unstable, in the sense that a trajectory started near this behavior will tend to move away from it.

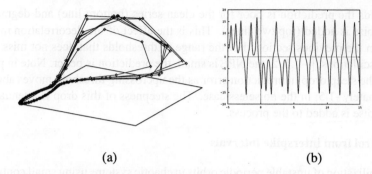

Figure 8: Subthreshold control of the unstable "figure-eight" orbit of the Lorenz attractor, using the interspike interval history only. (a) A plot in reconstructed state space. The three dimensions of the space are three consecutive time intervals. After an initial transient, the systems becomes trapped in the figure-eight orbit. (b) A plot of the time series from the Lorenz attractor, which is being integrated to form the spikes. The vertical line shows the time at which the control protocol is turned on.

In this feasibility study we used the Lorenz equations to generate an input signal $S(t) = x(t) + 25$. The parameters $\sigma = 28$ and $\rho = 10$ are set at the standard values, and the parameter β is nominally set at the standard value $8/3$, but will be used as control parameter.

Using the reconstruction of Lorenz dynamics solely from the interspike intervals, local linear models for a pair of points along the figure-eight orbit were constructed numerically. This step involves careful use of interpolation, since as Fig. 8 shows, any given point on the reconstructed orbit may be reached only rarely by an interspike interval vector.

Figure 8(a) shows the result of the control procedure. Using small perturbations to the nominal value of β, after an initial transient (the single inside loop), the stabilization is achieved, and the system will continue the figure-eight behavior indefinitely. A time trace of the x-coordinate of the underlying Lorenz attractor being used to drive the integrate-and-fire model is shown in Fig. 8(b). It was not used in the controlling process, and is shown here simply to verify success of the control procedure.

Superthreshold control of spike train dynamics is conceptually simpler. In this scenario, the goal is to change an erratic firing pattern to a regular pattern. The controller has the capability of causing the system to fire at any time, say through a large pulse applied to the system. We choose a desired fixed interspike interval and then ask the controller to apply superthreshold spikes to regulate the system to spiking rhythmically at that interval. In the on-demand pacing protocol, the controller applies the external spike whenever the interspike interval has exceeded the desired value. Obviously, this will cause the spike intervals to be capped at the desired level. What we will show is that in addition, under realistic assumptions on the dynamics, the spike times will be bounded from below as well, resulting in an evenly-timed spike sequence.

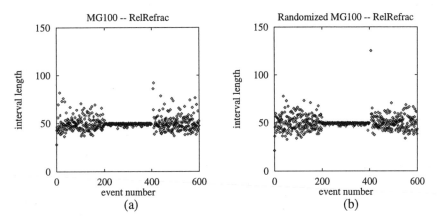

Figure 9: Application of superthreshold pulses to control an integrate-and-fire model with relative refractoriness. (a) The input signal to the integrator is a Mackey-Glass time series. Intervals are determined according to (1), where Θ is represented by a leaky potential. On-demand pacing control is applied between event numbers 200 and 400 only. (b) Input signal to the integrator is random noise with the same power spectrum as in (a). Similar control can be achieved for this random spike train.

To make a model which is perhaps closer to a neurophysiological process, we add to the integrate-and-fire model (1) some relative refractoriness. For this experiment we use a time series from the Mackey-Glass equation as the signal to be integrated. The (Mackey & Glass, 1977) delay differential equation is

$$\dot{x}(t) = -0.1x(t) + \frac{0.2x(t-\Delta)}{1 + x(t-\Delta)^{10}}. \tag{5}$$

When $\Delta = 30$, the attractor for this system has dimension around 3.5, according to numerical estimation. For $\Delta = 100$, the correlation dimension is approximately 7.1.

The relative refractoriness is included by changing the constant threshold Θ in the integrate-and-fire hypothesis (1) to a simple "leaky potential" function. The threshold Θ will decrease at a constant rate ($\dot{\Theta} = -0.5$ in our example) until firing, at which time Θ is replaced by $\Theta + 50$, representing the repolarization stage.

The results of control are shown in Fig. 9(a). A signal from Eq. (5) with $\Delta = 100$ is integrated until the threshold is met as in (1), with the assumption that the threshold Θ evolves as discussed above. For spike (event) numbers 1 to 200, the interspike interval lengths of the free-running system is graphed on the vertical axis. Since the integrated signal is 7-dimensional, no pattern can be ascertained by visual inspection. After event 200, the on-demand control protocol is turned on. The system immediately relaxes into a rhythmical behavior, with interspike intervals held near a fixed value. Note that not only are there no intervals above the nominal value (which is trivially enforced by on-demand pacing), but intervals below the nominal value have been largely eliminated as well. After spike 400, when the on-demand pacing is suspended, the system re-establishes its prior erratic behavior.

The same effect can be seen when the deterministic Mackey-Glass input is replaced by a random noise input. We manufacture the noise by creating a random-phase surrogate of the Mackey-Glass time series. That is, the phases of a Fourier-transformed version of the time series are randomized, and then the inverse Fourier transform is applied, resulting in a random signal with the same spectral characteristics as the deterministic Mackey-Glass signal used above. Figure 9(b) shows the same on-demand control technique as applied above. The control has an effect similar to that in the deterministic case.

6. Conclusions

We have suggested methods of studying and exploiting interspike interval series recorded from experiment, in the absence of known model equations. If they are generated according to hypothesis (1), the ISI vectors can be used to represent states of the underlying attractor in a similar manner as delay coordinate vectors from amplitude time series.

One strong motivation for this work is to advance understanding of information processing in neural systems. The possibility exists that much of the communication and processing that transpires in the central nervous system, for example, is coded in the temporal firing patterns of neural circuits. Our approach is designed to provide fundamental knowledge toward solving the so-called temporal coding problem.

This study shows the theoretical possibility of using the OGY control procedure with obvious modifications for systems measured only by spike trains. In systems where deterministic but aperiodic spike trains are measured, and where a global parameter is accessible for small time-dependent perturbations, this technique may be used to entrain the system into periodic spiking behavior.

Moreover, it was shown that the on-demand control procedure can be used to enforce equilibrium spiking behavior in a neurophysiologically motivated spiking model system. This capability was shown to exist for a deterministic system as well as a similar system that was principally random. We conclude that ability to control is not necessarily a reliable diagnostic for deterministic dynamical mechanisms.

Real experimental systems are certain to be much more complex in detail, and possibly not well represented by either integrate and fire or threshold crossing mechanisms. Further work with more realistic IF and TC models for specifics contexts is needed. We showed that dimension calculations and predictability from ISIs in practice are highly dependent on the details of the firing threshold. In a typical experiment, lack of control over this parameter could conceivably make accurate dimension estimation and prediction extremely difficult.

However, the basic message of these results is that under the right conditions, it is possible for a sequence of interevent interval to carry multidimensional state space information, in the same way that a time series can carry such information. Furthermore, as shown above, this capability degrades gracefully in the presence of noise.

Acknowledgements

The research of T.S. was supported in part by the U.S. National Science Foundation (Computational Mathematics program).

References

Castro, R., T. Sauer (1997) "Correlation dimension of attractors through interspike intervals", *Phys. Rev. E* 55:287.

Ding, M., C. Grebogi, E. Ott, T. Sauer, J.A. Yorke (1993) "Estimating correlation dimension from a chaotic time series: when does plateau occur?", *Physica D* 69:404.

Farmer, J.D., J. Sidorowich (1988) "Exploiting chaos to predict the future and reduce noise", *Evolution, Learning and Cognition*, ed. Y.C. Lee , World Scientific, Singapore.

Grassberger, P. (1990) "An optimized box-assisted algorithm for fractal dimension", *Phys. Rev. A* 148:63.

Grassberger, P., I. Procaccia (1983) "Measuring the strangeness of strange attractors", *Physica D* 9:189.

Lorenz, E. (1963) "Deterministic non-periodic flow", *J. Atmos. Sci.* 20:131-141.

Mackey, M.C., L. Glass (1977) "Oscillations and chaos in physiological control systems", *Science* 197:287.

Ott, E., C. Grebogi, E. Ott (1990) "Controlling chaos", *Phys. Rev. Lett.* 64:1196-1199.

Ott, E., T. Sauer, J.A. Yorke (1994) *Coping with Chaos: Analysis of Chaotic Data and the Exploitation of Chaotic Systems.* Wiley Interscience, New York, 418 pp.

Packard, N., J. Crutchfield, J.D. Farmer, and R. Shaw (1980) "Geometry from a time series", *Phys. Rev. Lett.* 45:712-715.

Preissl, H., A. Aersten, and G. Palm (1990) "Are fractal dimension a good measure for neuronal activity?", in: *Parallel Processing in Neural Systems and Computers*, R. Eckmiller, G. Hartman and G. Hauske (Eds.) Elsevier Science Publishers.

Rössler, O.E. (1976) "An equation for continuous chaos", *Phys. Lett.* 57A:397.

Sauer, T., J.A. Yorke, M. Casdagli (1991) "Embedology", *J. Stat. Phys.* 65:579-616.

Sauer, T. (1994) "Reconstruction of dynamical systems from interspike intervals", *Phys. Rev. Lett.* 24:3811.

Sauer, T. (1996) "Reconstruction of integrate-and-fire dynamics", in: *Nonlinear Dynamics and Time Series*, eds. C. Cutler, D. Kaplan, Fields Institute Publications, American Mathematical Society, Providence, RI.

Takens, F. (1981) "Detecting strange attactors in turbulence", in: *Dynamical Systems and Turbulence, Lecture Notes in Math.* 898, Springer-Verlag.

Takens, F. (1985) "On the numerical determination of the dimension of an attractor", *Lecture Notes in Math.* 1125, Springer-Verlag.

Theiler, J., S. Eubank, A. Longtin, B. Galdrakian, and J.D. Farmer (1992) "Testing for nonlinearity in time series: The method of surrogate data", *Physica D* (Amsterdam) 58:77.

H. Tuckwell (1988) *Introduction to Theoretical Neurobiology.* (Cambridge Univ. Press, Cambridge), Vols. 1 and 2.

SURROGATE DATA FOR NEURAL ACTIVITY

DONATELLA PETRACCHI
Istituto di Biofisica del CNR, Via S. Lorenzo 26,56127-Pisa (Italy)
e-mail: petracch@ib.pi.cnr.it

ABSTRACT

The search for surrogate data, that preserve the general statistical structure of the original signal, is a crucial task in searching for chaos in neural or cardiac activity. Here the method often used to generate surrogate data of spike trains is discussed, showing how it is inadequate. A new approach which preserve the short term correlation of the discharge is presented and applied to a neural-like discharge generated by a simple operational neural model. The difference in the conclusions about the low dimensionality of an irregular signal are highlighted.

1. Introduction.

Very often neural activity is quite irregular. This irregularity can be generated by the presence of many independent inputs and thus could be considered as due to noise, or altenatively can originate from intrinsic features of the neural encoding. The study of non linear predictability (Sugihara and May, 1990) (NLP) and the search of particular sequences (Schiff *et al.*, 1994) which are the sign of unstable periodicities (PUO, periodic unstable orbits) are the keys of two methods often used in recent years to detect dynamic features in neural encoding (see in this volume Barbi *et al.*, Moss *et al.*, and the references in their papers). In both methods it is necessary to compare the results of the analysis with those obtained by analysing surrogate data, which share with the original ones some basic statistical features. Evidence for dynamic components in neural encoding is obtained only when the results from the original data differ enough from those produced by surrogates.

Methods not based on the comparison with surrogates have also been used in the analysis of biological signals (Mueller-Gerking *et al.*, 1996). It could also be thought that the success of strategies aiming to control chaos in neural signals (for a description see in this volume Spano and Visarath) is in itself a proof that the signal is chaotic, but it has been proved that this conclusion is incorrect (Cristini and Collins, 1995, 1996).

As a matter of fact surrogate data can be easily used in association with any discriminating statistics, including Lyapunov coefficients and correlation dimensions, and for a correct interpretation of the behaviour of a biological system, the criteria adopted and the methods used to obtain surrogate data are critical.

For a continuous signal $s(t)$ the methods adopted aim to maintain its amplitude distribution and its spectrum (Osborne *et al.*, 1986; Theiler *et al.*, 1992; Chang *et al.*, 1995). It is easy to maintain the spectrum of a signal, by simply randomizing the phases in the Fourier transform: the new signal $s'(t)$ obtained by the inverse Fourier transform will have the same spectrum as the original one. However, in general $s'(t)$ will not have the same amplitude distribution as $s(t)$. A quite complicated method to maintain both the spectrum and the amplitude distribution is *gaussian scaling*. The first step is to replace the original signal $s(t)$ by a signal $h(t)$ with gaussian amplitudes (the substitution is made by ranking both the original data and the gaussian numbers and replacing the original data with gaussian numbers according to the rank). By subsequently randomizing the phases of $h(t)$, a gaussian signal $g(t)$ is obtained, and eventually the gaussian values of $g(t)$ are replaced by the original ones of corresponding rank; in this way the surrogate signal $s'(t)$ is generated.

The theoretical basis for this complicated procedure is clear: it simply tests whether the irregular signal comes out of a non-linear, non-dynamic transformation of a gaussian noise, namely whether $s(t)=f(g(t))$. The function f is experimentally obtained by the ranking procedure, and an explicitely formulated hypothesis for this method is that f be a graded invertible function.

There is instead no theoretical basis to apply the same procedure to a neural discharge, recorded as an all-or-none signal (f, if it exists, cannot be a graded invertible function). Nevertheless, the idea of maintaining the spectrum has been generally accepted, together with the requirement that the distribution of the intervals between spikes should also be maintained. The way in which this has been done is however quite strange. The discharge is not considered as a function of time; typically the value of the successive intervals T_i are plotted as function of i, the number used to label the intervals in the sequence according to their position. Let us call i the *order-label*; the order labeled sequence is also called tachogram[1].

How much unphysical is the order-label becomes evident by thinking to a case in which a sinusoidal stimulus of frequency f_0 produces a discharge with short intervals near the maxima of the sinusoid and long intervals near the minima; if the number of intervals near the maxima is not fixed (sometimes one, sometimes two or three) it is clear, intuitively, that the periodicity of the discharge is lost in the tachogram. However, this kind of plot has been largely used, in particular for cardiac activity. It is simple to draw such a plot and in some cases, for a quite regular activity, it can look similar to the real time course. But its use in searching for dynamic features of neural activity is arbitrary and the efforts made to obtain surrogate data maintaining the spectrum in the fictitious order-label i are not based on any physical requirement.

In this paper I will compare the real spectra in time of a neural-like discharge and of its label-based surrogates, showing that maintaining the "label spectrum" does not maintain the real spectrum. Then a method to maintain both the real spectrum and the interval distribution of a neural discharge will be proposed, and the reasons for maintaining the short-range correlation of the discharge in surrogate data will be

---

[1] Thinking to the etymology of this word, it does not seem very appropriate; however...... the use makes the meaning.

qualitatively discussed. Moreover, the search for PUO will be applied to a neural-like sequence of pulses and to three different kinds of surrogate data: those which preserve only the interval histogram, those based on the gaussian scaling of intervals considered as function of the order-label and those which preserve the real spectrum and the histogram of intervals.

2. Criteria for surrogate data.

2.1. The power spectrum of a neural discharge is not maintained by gaussian scaling of intervals in the tachogram.

This section is dedicated to show simulations in which the spectrum of a neural-like discharge is compared with that of surrogate data obtained by the gaussian scaling of intervals in the tachogram. This kind of surrogates will be called *G* surrogates.

To simulate a neural discharge, a sequence of pulses is obtained by software implementation of an integrate-and-fire system (I&F). A pulse is fired each time the time integral of a stimulus $V(t)=V_0(1+M\sin(2\pi\nu t))$ reaches the threshold value C; then the integral is reset to zero and integration starts again. The output of a simulation is a time function *s(t)* with possible values 0 or 1. This very simple model was used many years ago to simulate neural activity. The power spectrum of this neural-like discharge was computed via the autocorrelation function of the all-

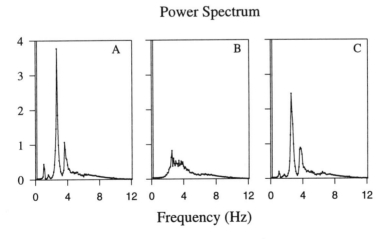

Fig. 1. A) Power spectrum (in arbitrary units) of the pulse train generated by the integrate-and-fire model. The input signal was $s(t) = s_o(1+M\cos(2\pi\nu t))$ and a Gaussian white noise with D = 1 was added to the stimulus signal. The frequency f_0 of the discharge in the absence of modulation was 2.6 pulses per second; the stimulus frequency ν was 1 Hz and the modulation depth M was 0.35. In B) and C) the power spectrum of the *G* and *A* surrogates are reported.

or-none signal. The square amplitude of the Fourier transform of the autocorrelation is reported in the plots of Fig. 1. Fig. 1A) shows the spectrum of the original signal: with the parameters used in this simulation the highest peak is around the average frequency of the discharge, since a peak at the same frequency is present even for M=0. Around it, at a distance which equals the stimulus frequency ν, two bands of different amplitude can be observed; these two bands are due to the stimulus-induced frequency modulation of the pulse train. The last peak at frequency ν is just due to the modulation of the average value of $s(t)$ produced by the stimulus. The spectrum of the *G* surrogate data is reported in Fig. 1B). Clearly the two spectra are different: surrogates obtained after the long procedure of gaussian scaling do not maintain the real spectrum of the original signal. Even the presence of a component at the stimulus frequency ν (due to the modulated input signal V(t)) is not visible in the spectrum of this kind of surrogate data. The same kind of spectrum is obtained by simply preserving the histogram of intervals, so that it seems that the only meaningful statistical feature that is maintained by gaussian scaling is the histogram of the intervals. In the next section a method will be introduced that approximately maintains the spectrum and the interval distribution of a neural discharge.

2.2. *How to maintain the power spectrum of a neural signal.*

The computation of the spectrum of a neural discharge via the Fourier transform of its autocorrelation function suggests a simple way to maintain the interval histogram and approximately the spectrum in the meantime. By definition, the autocorrelation function of a signal $s(t)$ is given by:

$$R(\tau) = \int s(t)s(t+\tau)dt \qquad (1)$$

and according to its definition it could be computed by multiplying s(t) by itself after a delay τ and summing up all the contributions. However s(t) is an all-or-none function; let be $H_1(t)$ the histogram of the intervals between two successive spikes, $H_2(t)$ the histogram of the intervals between each spike *j* and spike *j*+2, and in general $H_m(t)$ the histogram of the intervals between spike *j* and spike *j+m*. It is clear that R(τ), according to eq. (1), can be obtained by summing all the histograms of any order:

$$R(\tau) = \sum_m H_m(\tau) \qquad (2)$$

Therefore the problem of maintaining the spectrum coincides with the problem of maintaining the histograms of any order. Now in practice the correlation in a discharge does not last a very long time and the influence on neighbouring spikes is limited to only few spikes. If this is the case one can proceed in such a way as to save the correlation with the previous k intervals. For the spectrum reported in Fig. 1C) this is done with $k = 3$. Actually the program that generates the surrogate sequence starts by selecting at random three successive intervals. The fourth one is selected at random from the original data in the subset of intervals that have the three preceding ones similar (with a given tolerance) to the first three already selected. Then the procedure is iterated, each time taking into account the three preceding intervals. In the following, this kind of surrogates will be called A surrogates, because they were obtained preserving the autocorrelation. This procedure preserves also the histogram of intervals as shown in Fig. 2, where the histogram of intervals of the sequence generated by the I&F model and of the A surrogates are reported. So both the histogram and the power spectrum are preserved quite well by the A surrogates.

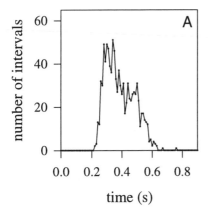 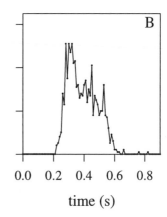

Fig. 2 - Histograms of intervals in the I&F data (A) and in the A surrogates (B)

2.3. PUOs in original data and in different kinds of surrogate data.

Two approaches (NLP and PUO) have been used quite often in the last few years for searching chaos or the signs of low-dimensional dynamics in spikes trains. The approach used here is that which searches for periodic unstable orbits (PUO) and is based on the identification in the data of particular episodes, approaching the unstable period. Plotting T_{i+1} vs. T_i such episodes will be characterised by sequences approaching and then departing from the 45° line, called encounters.

The comparison between the number of encounters found in the I&F discharge and in the different kinds of surrogates is reported in Table 1. Three kinds of surrogates will be considered: those obtained by gaussian scaling of intervals (*G* surrogates), those which simply maintain the interval distribution and are obtained directly by shuffling the original intervals (*I* surrogates) and the above-defined *A* surrogates, obtained by maintaining the autocorrelation of the discharge.

The encounters are defined by the same criteria used in previously published papers (Xing and Moss, 1996): three points approaching the line of identity followed by three points departing from it, with the same constraints on the slope and intersections of the lines fitting the two sets of approaching and departing points.

TABLE 1 - Number of encounters in original and surrogate data.

I&F data	Surrogate data		
	G	*I*	*A*
125	30	32	124
122	28	25	143
112	29	30	138
135	35	36	125
126	27	28	124

The search for encounters in the original and the *A* surrogate sequences gives the same order of encounters, correctly telling that no signs of low-dimensional dynamics are detected. This is in agreement with previous results (Petracchi, 1997), obtained by maintaining the correlation of the discharge with the stimulus, but is completely different from the result obtained by maintaining the "spectrum" of the tachogram. In fact, in the latter case the I&F is classified as a dynamic system with periodic unstable orbits. The same occurs if only the interval histogram is maintained. These results show how crucial are the criteria used to generate surrogate data in the search of chaos in spike trains.

3. Discussion

Two points will be discussed in turn: 1) the dependence of the "diagnosis" of low-dimensional dynamics in spike trains on the criteria used to generate surrogate data and 2) the reasons for maintaining the correlation between successive intervals in surrogate data.

3.1. Low-dimensional dynamics and the criteria used to generate surrogate data.

From Table 1 it is clear that surrogate data obtained preserving only the interval histogram give a false answer in searching low-dimensional dynamics in the neural-like discharge of the I&F. In fact, the number of encounters found in I surrogates is much lower than that found in the original discharge. Exactly the same situation occurs using data obtained by gaussian scaling of intervals in the tachogram: G surrogates give the same number of encounters as I surrogates. The reason for this is that the only meaningful statistical feature that is maintained after all the steps required for the gaussian scaling seems to be the histogram of intervals, as shown in Fig. 1, A) and B), where the power spectrum of the original I&F discharge and that of the G surrogates are reported. Therefore the first conclusion is that I surrogates and G surrogates appear to be equivalent and the work required for obtaining G surrogates can be avoided.

Moreover, the answer to the question whether signs of low-dimensional dynamics can be detected in the I&F discharge turns out to be negative by using surrogates that maintain the real spectrum of the original discharge. Which is the correct answer, the positive or the negative one? Clearly, by the definition of the I&F model used to generate the original data, the occurrence of a spike depends only on the values of the input since the last spike, which is enough to exclude the existence of unstable attractors or saddle points. Thus the correct answer is that given by the A surrogates, those obtained by maintaining the autocorrelation and the real power spectrum. The other two kinds of surrogates, G and I, obtained by gaussian scaling in the order-label or more simply by shuffling the intervals - the two procedures being equivalent - result to be useless in validating analyses aiming to detect the signature of dynamic systems in spiky signals.

3.2. Reason for maintaining the short-range correlation of the discharge in surrogate data.

The simple consideration of the starting point in the methods used to search for chaos in spike sequences suggests to maintain the correlation between neighbour intervals. The search for PUO is usually done in plots in which the interval T_i is reported on the abscissa while T_{i+1} is reported on the ordinate. The same kind of plot (often in more dimensions) is used also in the study of non-linear predictability. In this kind of plot the entire set of data is represented by a cloud of points.

Now it makes sense to require that the general shape of the cloud representing the signal should be maintained by the surrogate data. The centre of mass of the cloud is determined by the mean value of the interspike intervals, while the area of the cloud increases with the standard deviation of the intervals; these geometric features are preserved if the interval histogram is preserved. Saving the correlation between successive intervals preserves other geometric features. Consider for instance the cloud in Fig. 3. It represents a discharge where short intervals are followed by long

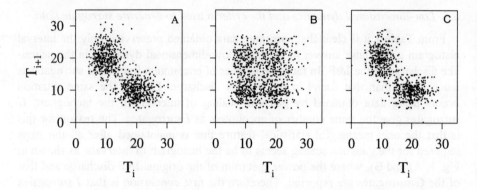

Fig. 3 - Intervals between spikes were extracted from two gaussian distributions centered on 10 and 20 (arbitrary units) respectively. In A) the original sequence, in B) surrogates preserving the intervals histogram, in C) surrogates that maintain the correlation between adjacent intervals, are reported.

ones and viceversa, so that the cloud appears as two separate ones (Fig. 3A). If only the interval histogram is preserved the cloud of the surrogate will be very different (four clouds are present in Fig. 3B). Instead maintaining the correlation between two successive intervals produces surrogate data that cannot be distiguished from the original at a first sight (Fig. 3C), and intuitively this is required from surrogate data - to look similar to the original ones.

References

Barbi M., S. Chillemi, A. Di Garbo, R. Balocchi, C. Carpeggiani, M. Emdin, C. Michelassi (1998) "Nonlinearity of normal sinus rhythm", *this volume*.

Chang T., T. Sauer, S. Schiff (1995) "Tests for nonlinearity in short stationary time series" *Chaos* 5:118-126.

Cristini D.J. and Collins J.J. (1995) "Controlling nonchaotic neuronal noise using chaos control techniques", *Physical Rev. Lett.*, 75:2782-2785.

Cristini D.J. and Collins J.J. (1996) "Using noise and chaos control to control nonchaotic systems, Physical Review E, 52:5806-5809.

Moss F., Pei X., Dolan K., Braun H., Huber M., Dewald M. "Low dimensional dynamics in sensory biology", *this volume*.

Muller-Gerking J., Martinerie J., Neuenschwander S., Pezard L., Renault B., Varela J.F. (1996) "Detecting non-linearities in neuro-electrical signals: a study of synchronous local field potentials", *Physica D*, 94: 65-91.

Osborne A.R., A.D. Kirwan, Provenzale A., Bergamasco L. (1986) "A search for chaotic behavior in large and mesoscale motions in the pacific ocean, *Physica D*: 23: 75-83.

Petracchi D. (1997) "The search for periodic unstable orbits in periodically driven spike trains", *Chaos, Solitons & Fractals* 8: 27-334.

Schiff S. J., Jerger K., Duong D. H., Chang T., Spano M. L., Ditto W. L.(1994) "Controlling chaos in the brain", *Nature* 370: 615-620.

Spano M.L. and Visarath I. "Controlling and maintaining chaos in biological system", *this volume*.

Sugihara, G. & R. May (1990) "Nonlinear forecasting as a way of distinguishing chaos from measurement error in a data series", *Nature* 344: 734-741.

Theiler, J., B. Galdrikian, A. Longtin, A. Eubank & J.D. Farmer (1992) "Testing for nonlinearity in time series: the method of surrogate data", *Physica* D 58: 77-94.

Xing P. and F. Moss. (1996) "Characterization of low-dimensional dynamics in the crayfish caudal photodetector", *Nature* 379: 618-621.

LOW DIMENSIONAL DYNAMICS IN SENSORY BIOLOGY

FRANK MOSS, XING PEI, KEVIN DOLAN
Center for Neurodynamics, University of Missouri at St. Louis
Saint Louis, MO 63121, USA
E-mail: mossf@umslvma.umsl.edu

and

HANS BRAUN, MARTIN HUBER, MATHIAS DEWALD
Department of Physiology, University of Marburg
Marburg, 35037, Germany

ABSTRACT

We describe a new method for detecting the presence of low-dimensional dynamical processes in noisy systems. Particular emphasis is placed on biological systems. The method is based on the statistical detection of rare events in a data file of time intervals. An algorithm is described which finds and counts the signatures of unstable periodic orbits in a heavily noise contaminated dynamical system. Such orbits are evidence of dissipative chaos. The method also detects limit cycles. Moreover, systems which are non stationary can be analyzed with success. Examples are given from neural recordings obtained from the photoreceptor in the crayfish 6th ganglion, catfish electroreceptor organs and cat cold receptor. In the latter case a quantitative Hodgkin-Huxley model is available and it is also analyzed using both the new and traditional methods.

1. Introduction

Low-dimensional dynamics is a term which describes dynamical behavior resulting from the nonlinear interaction of only a few variables. A notorious example is *chaos* where a nonlinear system with a minimum of three degrees of freedom can exhibit the classical signatures of this interesting behavior: exponential sensitivity to initial conditions and non integer, or fractal, dimensions of the motion in its phase space (Strogatz, 1994). In the early 1980s Predrag Cvitanovic introduced the first theory of the *structure* of dissipative chaos: a countable infinity of *un*stable periodic orbits (UPOs) (Cvitanovic, 1988; 1991; Artruso, *et al.*, 1990a; 1990b). This led to a new experimental characterization of chaos, which is, however, only now beginning to be exploited. Hopefully, one could detect the first few, lower order, orbits and measure the eigenvectors associated with the linearized motion near the unstable point on the orbit. The unstable point (in two dimensions) lies on the intersection of the stable and unstable manifolds. Straight line tangencies to the orbital trajectories near the unstable point define the eigenvectors. The eigenvalues are given by the inverse slopes of the tangent lines. Thus the detection of a set of such orbits and measures of the associated eigenvectors constitutes a characterization of chaos. A UPO occurs when a trajectory encounters a saddle potential as shown by the example in Fig. 1.

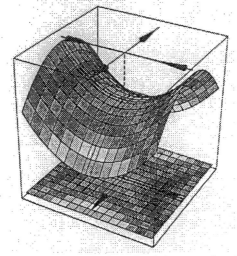

Fig. 1. A saddle potential. The stable (inward pointing arrows) and unstable (outward pointing arrows) are shown. The unstable point is at their intersection. Trajectories encountering a saddle are attracted toward the unstable point along the stable directions and repelled from it along the unstable directions. The structure of dissipative chaos is such that there is a countable infinity of orbits encountering such potentials.

It turns out that this method may be uniquely suited to the characterization of chaos in biological systems (Pierson and Moss, 1995). Such systems are notoriously non-stationary and noisy. In other words, living systems do not remain constant in time, and they are also subject to largely random processes. In this work, we assume that if a biological system is chaotic, we may detect the first few, lowest order, UPOs, but we will have to extract them from the noise. This means that a quantitative statistical method properly based on a null hypothesis will be necessary. Nonstatistical versions of this method have been applied to a number of chaos control problems (Ott, *et al.*, 1990; Hunt, *et al.*, 1991; Ditto and Pecora, 1993; Garfinkel, et al., 1992; Witkowski, *et al.*, 1995; Schiff, *et al.*, 1994a ; Roy, *et al.*, 1992; Rollins, *et al.*, 1993; Petrov, *et al.*, 1993), though in the case of the (noisy!) biological applications, a question has been raised. Exactly what is being controlled: noisy chaos or just plain noise (Christiani and Collins, 1995). Our search through a long biological file will be for *a few* events, or encounters, as we will call them, of the low order orbits with their unstable points (Kaplan, 1994).

These few encounters will nevertheless satisfy certain conditions which unmistakably signal the presence of an UPO. Even though rare, the encounters will satisfy a rigorous test of their statistical significance. Thus, as we shall show, even in short files which are quite noisy, we will be able to detect the presence of low-dimensional behavior with some quantifiable statistical accuracy (Pierson and Moss, 1995). Since we are able to deal with short files, we will be able to treat files resulting from non stationary conditions.

The method is based on the two different topologies which appear in the phase plane. The UPOs approach the unstable periodic point preferentially along the stable direction and then depart along the unstable manifold. The signature of this motion constitutes the topology of the UPO. By contrast, limit cycles, or stable periodic orbits (SPOs), approach their periodic points from all directions, and the signature of this motion represents the SPO. Our algorithm is designed to detect the signatures of UPOs in a data file. Our data are always displayed as a first return map of the time intervals T, between neural action potentials, that is, a plot of T_{n+1} versus T_n. An Example is shown in Fig. 2 (a).

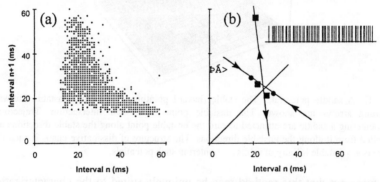

Fig. 2. (a) The first return map of a set of actual biological data. Hidden within the map are certain *temporal sequences* of points. (b) An *encounter* with an unstable point The temporal sequence shows three points which sequentially approach the unstable periodic point (the point on the 45° line closest to the intersection of the two lines) along the stable direction (inward pointing arrows) (circles) followed immediately by three which depart along the unstable direction (outward pointing arrows) (squares). One point (the square closest to the 45° line) is shared by the two sequences. The inset shows a segment of the neural signal within which the encounter was found. Many encounters may be found in a file such as shown in (a).

A periodic point is, by definition, the point for which $T_{n+1} \equiv T_n$. Thus the signature of the UPO is a *temporal sequence* of points in the phase plane, which first approach the periodic point along the stable direction and then depart from it along the unstable direction. Such a sequence is an encounter.

We then apply the method to the analysis of data from the crayfish caudal photoreceptor cell (CPR) (Pei and Moss, 1996 ; Pei, Wilkens and Moss, 1996).

This is a cell which is embedded in the 6th ganglion located in the tail of the crayfish (Wilkens, 1988; Hermann and Olsen, 1967). The ganglion receives input from an array of approximately 250 hair receptors located on the tail fan which respond to periodic water motions (Wilkens and Douglass, 1994) from approaching predators. The CPR has an internal pacemaker (oscillator). Thus periodically stimulating the hairs with water motions constitutes periodic forcing of a (surely nonlinear) oscillator, whence the possibility of chaotic and/or limit cycle behaviors. The light intensity falling on the CPR is used to change the "gain" of the internal oscillator. We have assembled data on chaotic behavior in this system as a function of external driving frequency and amplitude and light intensity on the CPR.

Next we turn to the catfish for further examples of the method, and show data from the its electroreceptor organs (Schäfer, *et al.*, 1990; Schäfer, *et al.*, 1995) These are temperature and electric field sensitive cells which can be spontaneously chaotic (Braun, *et al.*, 1997a). Bifurcations between limit cycles and UPOs are stimulated with variable external temperature.

Then, we show some data from the cat cold receptor as well as a Hodgkin-Huxley (HH) model of that modality (Braun, *et al.*, 1997b). The noise free HH model is chaotic in a certain region, as we show from the positive Lyapunov exponent. With added noise, so that it accurately mimics the cold receptor data, we show that the UPO method can easily detect chaos in the same region. Bifurcations between regions of SPOs and UPOs are easily discerned.

Finally, we add a brief discussion wherein the method of UPOs is contrasted with the traditional methods which determine the fractal dimension, the Lyapunov exponents or the linear or nonlinear predictability.

2. The Method

The method is essentially statistical. For applications in biology this is necessary, since most if not all biological modalities are dominated by noise, or random neural firings (Gerstein and Mandelbrot, 1964; Olsen and Schaffer, 1990). Thus our task is to extract the signatures of low-dimensional behavior from a file which is largely noise. To be convincing, this must be accomplished with quantifiable statistical accuracy. It is therefore, necessary to construct a null hypothesis for comparison with the data files. Files which play the role of the null hypothesis are called *surrogates* (Theiler, *et al.*, 1992) the signatures we seek are temporal sequences of points in the return maps, for example, shown in Fig. 2. They are thus characterized by short-term correlations between (only six) sequential neural firings. We can destroy that correlation by randomly shuffling the locations of the time intervals in the file. Thus a surrogate file is created from the original data.

It is important to note that a certain number of encounters N, will be found in the data file using a chosen set of criteria as the definition of an encounter. Using the *same* defining criteria, a (probably) different number N_S, will be found in the surrogate file. If statistically significant signatures exist in the data file, it will be found that $N > N_S$. The encounters found in the surrogate file can be considered to be "accidentals", that is, those which would be found only by chance in a random file. Because of this, it is necessary to find the *average* number of encounters in the surrogate files. In order to do this, one must construct many surrogates from the original data file, testing each for N_S, thus obtaining the mean $\langle N_S \rangle$ and standard deviation σ. A statistical measure of the presence of signatures in the data file against those in the surrogates is then,

$$K = \frac{N - \langle N_s \rangle}{\sigma}, \qquad (1)$$

Thus K measures the distance from the findings in the actual data file to those expected in the real file in units of the standard deviation σ. The latter quantity measures the "noise" intensity of the surrogate files. This is a standard statistical measure, and it is independent of the statistics of the process. The only requirement is that the process must have a finite first moment. However, if the process is Gaussian, then K can be directly related to a confidence level. For example, $K > 2$ indicates detection of the presence of UPOs with a confidence level greater than 95% (Bevington, 1969).

The defining criteria are arbitrary, but must select out of the data signatures based on the phase plane topology of the UPOs. We used the following to define an encounter:

1) Three time intervals which, in the phase plane, approach the 45° line at sequentially decreasing perpendicular distances, followed immediately by three points which depart at sequentially increasing perpendicular distances.

2) Straight lines are matched (by linear least squares fits) to the approaching three points and to the departing three points, and the following slope conditions applied: $0 \geq m_s > -1$ to the line of approaching points, and $-1 \geq m_{us} > -\infty$ to the line of departing points. These lines define the stable and unstable eigendirections and the inverse slopes define the eigenvalues of the linearized motion near the unstable periodic point. All biologically originated UPOs so far detected experimentally have stable and unstable eigenvectors with directions in the two defined ranges.

3) The perpendicular distance of the point of intersection of the two lines to the 45° line, ε, is smaller than half the mean of the three approach distances.

A sequence of 5 points on the phase plane (one interval is common to the approach and departure sets) which satisfy these three criteria is defined as an *encounter*.

It is interesting and important to note that this method, based as it is on phase plane topology, also detects the presence of limit cycles or SPOs. In this case it often happens for not too large noise that $K < 0$. This comes about in the following way. The algorithm searches for signatures of UPOs which have a specific phase plane topology: approach along a stable direction followed by departure along an unstable direction. But SPOs have an entirely different topology: all directions are stable. Thus it may happen that only a very small number (or even zero) signatures (of UPOs) are found in the data file, which contain a certain order. But the surrogates are more disordered, so that there are more chance satisfactions of the criteria than in the data file. Thus $N < N_s$, resulting in negative K. It turns out that this property is very useful.

A simplified algorithm, which runs faster, and hence is suitable for use on line during experiments, omits the calculation of surrogate files. For on line applications one can simply break the incoming time interval data into segments, of say 200 or 300 time intervals, and find the number of signatures N, in each segment. Then the percentage of points comprising encounters which satisfy the selection criteria is computed and continuously displayed. We have found with numerical simulations and a simple statistical theory that the percentage of such points in random (and hence surrogate) files is about 3.6%, and this is approximately independent of the distribution of the process. Thus, on line, whenever the displayed percentage jumps significantly above this level, say to 5 or 6 %, there is a strong indication of the presence of UPOs.

An alternative method has also recently been developed (So, et. al., 1996). Though it is mathematically more sophisticated, it is also more computationally intensive.

3. UPOs in the Crayfish 6th Ganglion.

The crayfish, *Procambarus clarkii* has two bilaterally located photoreceptor cells embedded in its 6th, or terminal, ganglion. They are connected synaptically to the interneurons of the ganglion, which are, in turn, connected to the sensory neurons which enervate the hydrodynamically sensitive hairs on the tail fan of the animal. The preparation is made by excising the entire tail two cms above the 6th ganglion and exposing both the sensory nerve cord and one output photoreceptor neuron. Electrodes and preamplifiers are attached to the neurons and light is introduced to the receptive area of the photoreceptor by a fiber optic cable. The preparation is shown in Fig. 3.

The steady state light intensity I, can be adjusted, and is a control parameter of the experiment. The hydrodynamic stimulus is applied to the complete hair array as water motion in the direction shown and consists of a sine wave of amplitude A and frequency f. Information about the hair motions is transmitted to the ganglion via the sensory neurons. The interneurons then transfer this information to the photoreceptor cell.

Two recording sites (electrodes and amplifiers) are shown. On the sensory neuron, one sees only noise plus information about the signal, but on the photoreceptor neuron, one sees complex behavior (Pei, Wilkens and Moss, 1996).

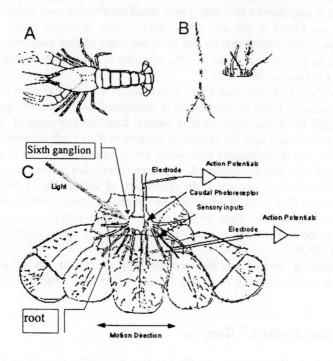

Fig. 3. The preparation, showing (A) the whole crayfish, (B) a single hair receptor and (C) the tailfan and 6th ganglion preparation.

The photoreceptor has a subthreshold embedded oscillator. Thus application of the single frequency stimulus constitutes periodic forcing of this nonlinear oscillator and hence the possibility of chaos. The light intensity controls the sensitivity of the signal transduction process. With recordings from the photoreceptor we can observe SPOs, or limit cycles, and UPOs depending on the parameters *I, A,* and *f*. Example data are shown in Figs. 4.

We note that UPOs are present for only certain ranges of parameters, as is always the case for periodically forced nonlinear oscillators.

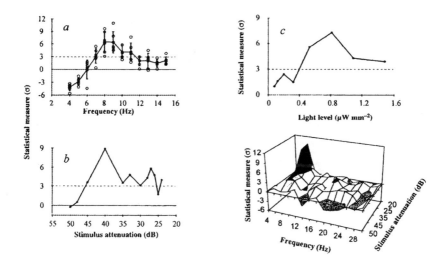

Fig. 4. The statistical measure K versus parameter values: (a) stimulus frequency; (b) stimulus amplitude; (c) light level. (d) shows a composite plot. Negative values of K indicate limit cycles. Strongly positive values ($K > 3$) indicate periodic unstable orbits, a signature of the structure of chaos.

4. Catfish Electroreceptors

Much work has been done on electroreceptive organs of both catfish and sharks (Schäfer, et al., 1990; Braun, et al., 1994; Schäfer, et al., 1995). These organs are ampullary receptors, see Fig. 5, and are multimodal, that is, they are sensitive both to electric fields and to the ambient temperature. Moreover, they are characterized by remarkably stable, subthreshold oscillators. The spiking patterns shown in wide ranges of temperature show the multipeaked interspike interval histograms characteristic of noise mediated subthreshold oscillators (Longtin, et al., 1991; Longtin, et al., 1994). We have found that in the catfish *Ictalurus nebulosus* these receptors spontaneously show UPOs, that is it is not necessary to force them with an external periodic stimulus. Our analyses are all using the temperature as the control parameter.

In Fig. 6 we show a temperature induced bifurcation such that the interval histogram is initially monomodal but bifurcates to bimodality sometime after the temperature change. UPOs are present before and during the bifurcation but not afterwards. The K values are shown above the bars in (c) and the length of the bar represents the time duration of the file over which the value of K was obtained. The step temperature change is shown in (e). Note the strong indication of chaos just before the bifurcation and the indication of a limit cycle well after.

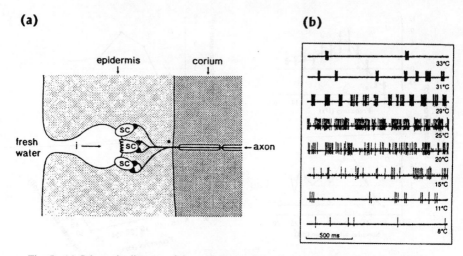

Fig. 5. (a) Schematic diagram of the catfish ampullary electroreceptor. The outer medium, fresh water, is at left, the inside of the fish is at right Recordings are made from the axon. The excitatory current is i. The sensory cells, where spike generation is assumed to take place are labeled SC. (b) Recorded spike trains at various temperatures.

5. The Cat Cold Receptor and a Hodgkin-Huxley Model

Subthreshold oscillators are also found in mammalian cold receptors (Braun, et al., 1997b, Longtin and Hinzer, 1996). These are sensory neurons which can be found in the skin and sense the ambient air temperature. The data herein presented are for cat, but similar data have been obtained for mice and vampire bats (Schäfer, et al., 1988). In Fig. 7(a) a temporal record of all time intervals recorded from the cat cold receptor are shown for a series of step changes in temperature from 35 to 10°C. In Fig. 7(c) is shown a similar result for a Hodgkin-Huxley model of this neuron with added noise. The parameters of the HH model are given in the Appendix, but the added noise intensity has been adjusted to give the best representation of the physiological data. The temperature steps are shown in (b). In (a) and (c) note that "skipping", which is evidence for the subthreshold noisy oscillator is evident at 35 °C. The remaining intervals until 10 °C show more-or-less complex sets of limit cycles and for these data our analyses show $K < 0$. The only segment of the data which is chaotic is the one at 10°C, and the one at 5°C on the far right in the simulation, though because of the noise this is not apparent in the Figure.

In Fig. 8(a) we show the noise free HH model over an expanded temperature range, 15 to 5 °C which includes the region of chaos in the physiological data. Without the noise, the chaotic region centered around 10 °C is quite evident, as shown by the positive Lyapunov exponents in (b). This exponent is zero in the

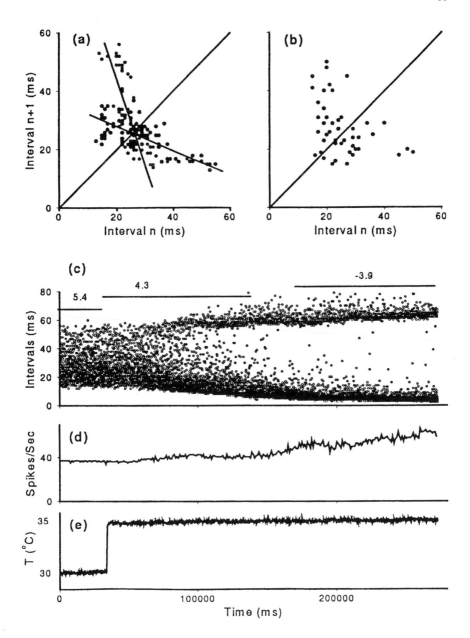

Fig. 6. (a) Several sets of approach and departure points with eigendirections shown by the lines for $K = 5.4$; (b) the surrogate of the data shown in (a), and (c) a plot of all the recorded intervals versus time, showing the bifurcation to bimodality. The temperature is shown in (e) and the mean discharge rate in (d).

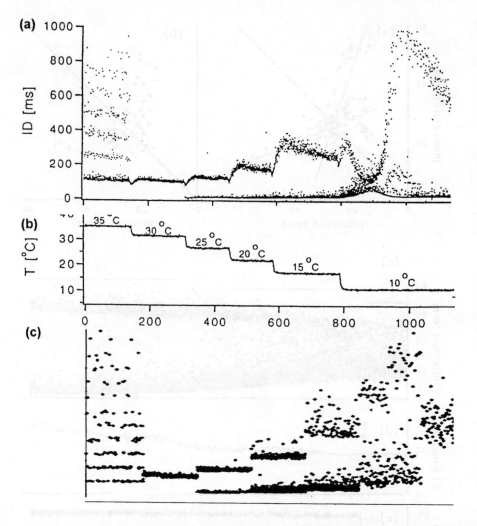

Fig. 7 (a) Physiological data from the cat cold receptor. ID are the time intervals between spikes. (b) The temperature steps for which the data in (a) were obtained. (c) The time intervals for the same range of temperature steps but with an extra step at 5 °C added on the far right.

region of limit cycles and positive in the chaotic region. In (c) we show the region with noise: the same noise intensity as used in Fig. 7 (c).

We note that the chaotic region is slightly shifted toward the higher temperatures by the noise and suppressed in the region around 10 °C. The measured

K values are shown in (d). They show that the chaotic region has been shifted by the noise to a narrower range around 13.5 °C.

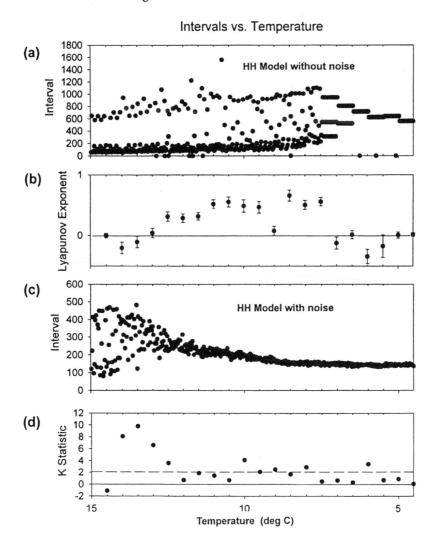

Fig 8. (a) Time intervals from the HH model without noise versus temperature. (b) The Lyapunov exponent of the data shown in (a) showing the chaotic region (positive exponents) centered around 10 °C. (c) The HH model with noise [as in Fig. 7 (c)] showing the collapse of the multiple limit cycles into a single cycle and the shift of complex behavior toward around 13 °C. (d) The K values for the data in (c) showing a strong indication of chaos in at 13 °C.

6. Discussion

It is useless to try to run any of the standard algorithms for detecting and characterizing chaos, for example the Lyapunov exponent (Wolf, *et al.*, 1985) or the correlation dimension (Grassberger and Procaccia, 1983), on data such as that shown in Fig. 7. The noise is large enough that unreliable results are obtained even for unrealistically long data sets. Repeated efforts to conclusively detect chaos in biological files using these algorithms have been largely inconclusive (Ruelle, 1994). Efforts to detect low dimensional nonlinear behavior using predictor methods have likewise been less than conclusive (Chang, *et al.*, 1994; Schiff, *et al.*, 1994b; Scott, *et al.*, 1995, Sugihara and May, 1990). In any case, predictor methods cannot distinguish between noisy SPOs and UPOs. Moreover, the data sets from biological experiments are usually short and non-stationary. Indeed, in Fig. 7, the only segment which we have found to be chaotic is the one at 10°C, and that segment is very non-stationary as indicated by the large positive going transient. Yet the method can easily detect the UPOs and measure their average eigenvectors as shown by Fig. 7 (d) and distinguish them from neighboring SPOs. Proof that the method is detecting chaos is offered by the noise free simulation of Fig. 8, where the positive Lyapunov exponent shown in (c) clearly identifies the chaotic region as the same one for which we saw large positive K values in the noisy HH simulation of Fig. 7 (c).

In conclusion, we have demonstrated a new statistically based method which can detect UPOs in noisy data and distinguish them from SPOs. The method has been demonstrated using experimental data from several biological sensory systems.

Acknowledgement

The research described in this article is supported by the U.S. Office of Naval Research, Physics Division.

Appendix

The numerical parameters used in the HH model are given below:

Membrane equation:

$$C_M \, dV/dt = -g_l (V - V_l) - I_d - I_r - I_{sd} - I_{sr} + gw \tag{A1}$$

Fast ionic currents which generate the action potentials:

$$I_d = \rho g_d \, m_{d\infty}(V - V_d); \quad m_{d\infty} = 1/(1 + \exp[-0.25(V+25)]) \tag{A2}$$

$$I_r = \rho g_r \, n_r(V - V_r); \quad dn_r/dt = \phi(n_{r\infty} - n_r)/\tau_r ; \tag{A3}$$

$n_{r\infty} = 1/(1 + \exp[-0.25(V+25)])$ (A4)

Slow ionic currents which generate oscillatory activity:

$I_{sd} = \rho g_{sd} m_{sd}(V - V_{sd});$ (A5)

$dm_{sd}/dt = \phi(m_{sd\infty} - m_{sd})/\tau_{sd};$ $m_{sd\infty} = 1/(1 + \exp[-0.09(V+40)])$ (A6)

$I_{sr} = \rho g_{sr} n_{sr}(V - V_{sr});$ $dn_{sr}/dt = \phi(-\eta I_{sd} - k n_{sr})/\tau_{sr}$ (A7)

Temperature Scaling:

Conductances: $\rho = 1.3^{(T-T_0)/10};$ $T_0 = 25\,°C$ (A8)

Time constants: $\varphi = 3.0^{(T-T_0)/10};$ $T_0 = 25\,°C$ (A9)

Noise:

gw is a Gaussian noise of standard deviation σ.

Numerical values of parameters:

Conductances: $g_{sd} = 0.25;$ $g_{sr} = 0.4;$ $g_d = 1.5;$ $g_r = 2.0$ mS/cm^2

Time constants: $\tau_{sd} = 10;$ $\tau_{sr} = 20;$ $\tau_r = 2.0$ ms

$\eta = 0.0012;$ $k = 0.17$

$V_{sd} = V_d = 50;$ $V_{sr} = V_r = -90;$ $V_l = -60$ mV

References

Artuso, R., E. Aurell and P. Cvitanovic (1990b) "Recycling of strange sets II: applications". *Nonlinearity* 3: 361-375.

Artuso, R., E. Aurell, and P. Cvitanovic (1990a) "Recycling of strange sets I: cycle expansions". *Nonlinearity* 3: 325-338.

Bevington, P.R. (1969) *Data Reduction and Error Analysis*, McGraw-Hill, New York, pp. 48-49.

Braun, H.A., H. Wissing, K. Schäfer, and M.C. Hirsch (1994) "Oscillation and noise determine signal transduction in shark multimodal sensory cells". *Nature* 367: 270-273.

Braun, H.A., K. Schafer, K. Voigt, R. Peters, F. Bretschneider, X. Pei, L. Wilkens and F. Moss (1997a) "Low-dimensional dynamics in sensory biology 1: thermally sensitive electroreceptors of the catfish". *J. Comp. Neurosci.*, in press.

Braun, H.A., M. Huber, M. Dewald, K. Schäfer and K. Voigt (1997b) "Computer simulations of neuronal signal transduction: the role of nonlinear dynamics and noise". *Intern. J. Bifurc. and Chaos*, in press.

Chang, T., S.J. Schiff, T. Sauer, J.-P. Gossard, and R.E. Burke (1994) "Stochastic versus deterministic variability in simple neuronal circuits. I. Monosynaptic and spinal cord reflexes". *Biophys. J.* 67: 671-683.

Christini, D.J. and J.J. Collins (1995) "Controlling nonchaotic neuronal noise using chaos control techniques". *Phys. Rev. Lett.* 75: 2782-2785.

Cvitanovic, P. (1988) "Invariant measurement of strange sets in terms of cycles". *Phys. Rev. Lett.* 61: 2729-2732.

Cvitanovic, P. (1991) "Periodic orbits as the skeleton of classical and quantum chaos". *Physica D* 51: 138-152.

Ditto, W. and L. Pecora (1993) "Mastering chaos". *Scientific American*, August 78-84.

Ditto, W.L., S.N. Rauseo, and M.L. Spano (1990) "Experimental control of chaos". *Phys. Rev. Lett.* 65: 3211-3214.

Garfinkel, A., M.L. Spano, W.L. Ditto and J.N. Weiss (1992) "Controlling cardiac chaos". *Science*, 257: 1230-1235.

Gerstein, G. and B. Mandelbrot (1964) "Random walk models for the spike activity of a single neuron". *Biophys. J.* 4: 41-68.

Grassberger, P. and I. Procaccia (1983) "On the characterization of strange attractors". *Phys. Rev. Lett.* 50: 346-349.

Hermann HT and Olsen R.E. (1967) "Dynamic statistics of crayfish caudal photoreceptors". *Biophysics J.* 7: 279-296.

Hunt, E.R. (1991) "Stabilizing high-period orbits in a chaotic system-the diode resonator". *Phys. Rev. Lett.* 67: 1953-1955.

Kaplan, D.T. (1994) "Exceptional events as evidence for determinism". *Physica D* 73: 38-48.

Longtin, A. and K. Hinzer (1996) "Encoding with bursting, subthreshold oscillations and noise in mammalian cold receptors". *Neural Comp.* 8: 217-257.

Longtin, A., A. Bulsara, and F. Moss (1991) "Time-interval sequences in bistable systems and the noise-induced transmission of information by sensory neurons". *Phys. Rev. Lett.* 67: 656-659.

Longtin, A., A. Bulsara, D. Pierson, and F. Moss (1994) "Bistability and the dynamics of periodically forced sensory neurons". *Biol. Cybern.* 70: 569-578.

Olsen, L.F. and W.M. Schaffer (1990) "Chaos versus noisy periodicity: alternative hypotheses for childhood epidemics". *Science* 249: 499-504.

Ott, E., C. Grebogi, and J.A. Yorke (1990) "Controlling chaos". *Phys. Rev. Lett.* 64: 1196-1199.

Pei, X and F. Moss (1996) "Characterization of low dimensional dynamics in the crayfish caudal photoreceptor". *Nature* 379: 618-621.

Pei, X., L. Wilkens and F. Moss (1996) "Light enhances hydrodynamic signaling in the multimodal caudal photoreceptor interneurons of the crayfish". *J. Neurophysiol.* 76: 3002-3011.

Petrov, V., V. Gaspar, J. Masere, and K. Showalter (1993) "Controlling chaos in the Belousov-Zhabotinsky reaction". *Nature* 361: 240-243.

Pierson, D. and F. Moss (1995) "Detecting periodic unstable points in noisy chaotic and limit cycle attractors with applications to biology". *Phys. Rev. Lett.* 75: 2124-2127.

Rollins, R.W., P. Parmananda, and P. Sherard (1993) "Controlling chaos in highly dissipative systems: A simple recursive algorithm". *Phys. Rev. E* . 47: R780-R784.

Roy, R., T.W. Murphy, T.D. Maier, Z. Gillis, and E.R. Hunt (1992) "Dynamical control of a chaotic laser: experimental stabilization of a globally coupled system". *Phys. Rev. Lett.* 68: 1259-1262.

Ruelle, D. (1994) "Where can one hope to profitably apply the ideas of chaos?". *Physics Today* 47 (July) 24-30.

Schäfer, K., H.A. Braun, R.C. Peters, and F. Bretschneider (1995) "Periodic firing pattern in afferent discharges from electroreceptor organs of catfish". *Pflügers Arch. Eur. J. Physiol.* 429: 378-385.

Schäfer, K., H.A. Braun, F. Bretschneider, P.F.M. Teunis, and R.C. Peters (1990) "Ampullary electroreceptors in catfish (Teleostei): temperature dependence of stimulus transduction". *Pflügers Arch. Eur. J. Physiol.* 417: 100-105.

Schäfer, K., H.A. Braun and L. Kürten (198 8) "Analysis of cold and warm receptor activity in vampire bats and mice". *Pflügers Arch. Eur. J. Physiol.* 412: 188-194.

Schiff, S.J., K. Jerger, D. Duong, T. Chang, M.L. Spano, and W.L. Ditto (1994a) "Controlling chaos in the brain". *Nature* 370: 615-620.

Schiff, S.J., K. Jerger, T. Chang, T. Sauer, and P.G. Aitken (1994b) "Stochastic versus deterministic variability in simple neuronal circuits. II. Hippocampal slice". *Biophys. J.* 67: 684-691.

Scott, D.A. and S. J. Schiff (1995) "Predictability of EEG Interictal Spikes". *Biophys. J.* 69: 1748-1757.

So, P., E. Ott, S.J. Schiff, D.T. Kaplan, T. Sauer, and C. Grebogi (1996) "Detecting unstable periodic orbits in chaotic experimental data". *Phys. Rev. Lett.* 76: 4705-4708.

Strogatz, S.H. (1994) *Nonlinear Dynamics and Chaos*, Addison-Wesley, Reading, MA.

Sugihara, G. and R.M. May (1990) "Nonlinear forecasting as a way of distinguishing chaos from measurement error in time series". *Nature* 344: 734-741.

Theiler, J., S. Eubank, A. Longtin, B. Galdrikian, and J.D. Farmer (1992) "Testing for nonlinearity in time series: the method of surrogate data". *Physica D* 58: 77-94.

Wilkens L.A. (1988) "The crayfish caudal photoreceptor: Advances and questions after the first half century". *Comp Biochem Physiol.* 91: 61-68.

Wilkens, L and J. Douglass (1994) "A stimulus paradigm for analysis of near-field hydrodynamic sensitivity in crustaceans". *J. Exp. Biol.* 189: 263-272.

Witkowski, F.X., K.M. Kavanagh, P.A. Penkoske, R. Plonsey, M. Spano, W.L. Ditto, and D.T. Kaplan (1995) "Evidence for detterminism in ventricular fibrillation". *Phys. Rev. Lett.* 75: 1230-1233.

Wolf, A., J.B. Swift, H.L. Swinney, and J.A.Vastano (1985) "Determining Lyapunov exponents from a time series". *Physica D* 16: 285-317.

FRACTAL ANALYSIS IN THE HEART

R. BALOCCHI

and

C. MICHELASSI, C. CARPEGGIANI, M. CASTELLARI, M.G. TRIVELLA

CNR, Institute of Clinical Physiology, via Trieste, 41
56125-Pisa, Italy
E-mail: balocchi@nsifc.ifc.pi.cnr.it

ABSTRACT

This paper describes the application of some concepts from spatial and temporal fractal analysis to the cardiovascular system. After a short introduction to the functioning of the heart and to important concepts of fractal geometry, two examples of temporal fractal properties of a process will be discussed by analysing the nature of the fluctuations of the heart rate variability in one normal and one transplanted heart. The spatial fractal properties of a process will be presented, introducing an example taken from a study on the coronary blood flow distribution in canine heart.

1. Introduction

Physiological systems are often characterised by being at the same time well organised but complex 'objects'. The term complexity can refer to the dynamical properties of a process or it can describe the features of a structure; in the first case it is usually investigated by using the tools of nonlinear analysis and chaos while in the second a suitable characterisation can be supplied by the concepts of fractal theory.

Of particular interest is the cardiovascular system, whose complexity has been widely investigated in these last years. Literature offers a great variety of investigations, ranging from modelling the dynamics of the electrical activation of the cardiac muscle (Giaquinta *et al*., 1996; Holden *et al*., 1996), to the analysis of nonlinear features in the heartbeat series (Sugihara & May, 1990; Kaplan, 1995; Di Garbo *et al.,* 1996), to the characterisation of the blood flow distribution in the cardiac muscle (Bassinghwaithe *et al*., 1989; Iversen & Nicolaysen, 1995; Trivella *et al*., 1996) or to the fractal properties of the heart rate variability (HRV) (Goldberger *et al*., 1990; Balocchi *et al*., 1994). With regard to HRV, it is important to recall that fractality is not a concept related only to the spatial characteristics of an object; fractal properties can pertain also to processes evolving in time.

In what follows some spatial and temporal fractal properties of the heart will be discussed. Next section illustrates, in a very elementary way, the physiology at the basis of the heart functioning, while section 3 introduces a short description of the principal concepts of fractal theory relevant for this lesson. These concepts are illustrated with examples in section 5, showing the temporal fractal properties of the human HRV and the fractal spatial distribution of the coronary blood flow in the canine cardiac muscle.

2. Functioning of the Heart

To help understanding the present lesson a short, simplified introduction to the heart functioning will be given. In particular, the heart will be described with respect to three major topics: 1) generation of the surface electrocardiogram, 2) heart rate variability, 3) myocardial perfusion.

2.1. Cardiac Activation and Control Mechanisms

The cardiac tissue possesses the property of spontaneously and rhythmically initiating its own beat.

In mammalian heart the natural pacemaker is the sinoatrial node (SA), located in the right atrium, from which the cardiac impulse spreads throughout both atria to the atrioventricular node (AV); this activation results in a contraction of the atria.

The AV nodal fibres gradually merge with the bundle of His which divides into the right and left ventricles and, in turn, subdivide into the Purkinje fibres. The high conductivity of these fibres permits the activation and the subsequent contraction of the ventricles. All this activity is aimed at collecting and pumping blood from and to the body districts, including, of course, the cardiac muscle itself, fed by the coronary tree.

The electrocardiogram (ECG) is the most common way to detect the cardiac electrical activity from the body surface. Figure 1 is a schematic representation of the ECG: the p wave corresponds to atria activation, QRS waves to ventricular activation, T wave to repolarization of ventricles. Then the next cardiac cycle occurs.

2.2. Heart Rate Variability

The time elapsed between two consecutive beats is well represented by the time interval separating the corresponding R waves (RR values): Fig. 2(a) is a sample of the RR time series of a healthy subject. The beat-to-beat variability

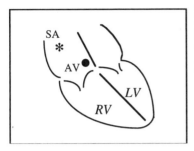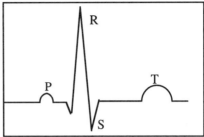

Figure 1. Schematic representation of the heart (left panel) and the ECG (right panel). The star and the black circle indicate the approximate location of the sinoatrial node (SA) and atrioventricular node (AV), respectively, in the right atrium; RV and LV indicate right and left ventricles respectively. See text for details on ECG.

present in the series is physiological and depends on the influence of many factors, central and local, nervous and chemical; the autonomic nervous system, however, plays the major role, with the competing activity of its sympathetic and parasympathetic branches that modulate the heart rate (Pagani *et al.*, 1986). To schematise, we could say that an increase in sympathetic activity produces an increase in HR, while an increase in parasympathetic activity results in a HR deceleration. The autonomic nervous system exerts its influence on the intact heart, but not anymore on the denervated transplanted heart; in transplanted subjects, however, the HRV is still present, even if dramatically reduced, as can be seen in Fig. 2(b). Therefore, intact and denervated hearts represent two extreme conditions as to the influence of the autonomic nervous system.

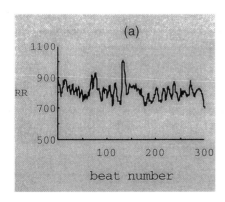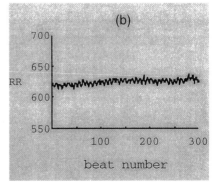

Figure 2. Interbeat time intervals (RR) versus beat number. Panel (a) refers to a healthy heart with its physiologic variability. Panel (b) refers to a transplanted heart: a dramatically reduced variability is clearly visible.

2.3 Myocardium Perfusion

The primary factors responsible for perfusion of the myocardium (the cardiac muscle) is the aortic pressure. The right and left coronary arteries, which arise from the aorta, provide the entire blood supply to the myocardium; they branch in vessels whose diameter becomes smaller and smaller.

As for HRV, neural, neurohumoral and metabolic factors influence the coronary flow, affecting, for instance, the contractility of the heart and the coronary resistance. It is known that the distribution of the blood flow into the cardiac muscle is not uniform: it shows a considerable spatial heterogeneity, which has been widely investigated in animals. In section 4 an example taken from laboratory experiments on canine hearts will be reported.

3. Fractal Geometry

3.1 Self-similarity

Many objects are characterised by having a specific spatial or temporal scale at which they are completely defined: these objects are well described by the concepts of the Euclidean geometry. On the other hand, there are objects that reveal always the same features at different levels of magnification, therefore lacking of a single characteristic scale; these objects are called 'fractals' and this property is known as 'self-similarity'.

Geometrical self-similarity is an abstract concept that does not apply to real life objects. Biological structures can reveal more and more details at different length scales, but they can't be defined geometrically self-similar. They can, however, retain the same statistical properties as the magnification scale changes, so being statistically self-similar. Statistical self-similarity can be expressed as

$$h^{-H}[x(t_i + h\Delta t) - x(t_i)] \stackrel{d}{=} [x(t_i + \Delta t) - x(t_i)] \tag{1}$$

where $\stackrel{d}{=}$ means equality in statistical distribution and H is the similarity exponent. The value of H can be seen as a measure of the 'scaling', that is of the dependence between the value measured and the scale of magnification.

3.2 Power law

Temporal self similarity is intrinsically related to the spectral behaviour. The characteristic of reproducing a 'scaled fluctuation' at each length scale produces, in the frequency domain, a power-law power spectrum of the form $p(f) = f^{-\alpha}$, with α a positive constant.

It is important to outline that self-similarity implies power-law spectrum but not vice versa (refer to the well known counterexample of the saw-tooth signal).

After a logarithmic transformation, the expression of power-law scaling results in a linear relation between frequencies and their powers $\log(p) = \alpha \log(f)$.

In a fractal series with power spectrum of the type $f^{-\alpha}$ and power spectrum of the increments of the type $f^{-\beta}$ the three exponents H, α and β are linked together by the relations: $\alpha = 2H+1$ and $\beta = 1-2H$. Furthermore the fractal dimension D_f and the dimension of the series graph D_g can be expressed according to: $D_f = 1/H$ and $D_g = 2 - H$.

3.3 Relative Dispersion

In analogy with the fractal dimension, which gives information on the complexity of an object, we can measure the degree of heterogeneity of the spatial distribution of a substance in a medium of a given size by measuring the Relative Dispersion RD=SD/mean. When the standard deviation is resolution-dependent, as it is for blood flow in cardiac tissue (Bassingthwaighte et al., 1989), RD becomes inappropriate to describe the overall heterogeneity of the substance. A better parameter to describe this feature is the fractal dimension D, which can be computed according to the equation

$$RD(m) = RD(m_{ref}).(m/m_{ref})^{1-D} \qquad (2)$$

where RD(m) indicates the relative dispersion of the substance at a given mass of the pieces of tissue and RD(m_{ref}) is the same quantity evaluated at an arbitrarily chosen reference mass m_{ref}.

4. Methods

4.1 Detrended Fluctuation Analysis

The basic idea beyond Detrended Fluctuation Analysis (DFA) (Peng et al., 1995) is to evaluate, and then remove, the local trend present in the interbeat intervals of the series B_i, prior to the analysis of its fluctuations. The procedure consists of the following steps: 1) create the new series $y(k) = \Sigma_{i=1,k} [B_i - B_{ave}]$, where B_{ave} are the series means, 2) create a grid of boxes of equal length n, 3) evaluate the local trend $y_n(k)$ as the least-square fit in each box, 4) evaluate the average deviation from the local trend according to $F(n) = \sqrt{\frac{1}{N} \sum_{k=1}^{N} [y(k) - y_n(k)]^2}$. In presence of scaling behaviour, $log(F(n))$ and $log(n)$ are linked by a linear relationship of the type $log(F(n)) = \alpha\, log(n)$. Specific values of α are associated to different dynamics: a value of $\alpha = 1$ is peculiar of $1/f$ power-law processes, while $\alpha = 0.5$ and $\alpha = 1.5$ indicate, respectively, white (uncorrelated) and Brownian (correlated) noise.

4.2 Data Collection

Analysis of human HRV. We present here two examples concerning the analysis of the fluctuations of HRV in one normal and one transplanted subject. Their ECG's, recorded with the Holter technique for 24 hours, were processed to detect the RR interbeat time intervals. The two series were then analysed using the DFA method to check for the existence of a temporal scaling behaviour.

Analysis of canine coronary blood flow distribution. The example reported here is taken from a study concerning the microvascular perfusion of canine myocardium (Trivella et al., 1996). In anaesthetised open chest dogs an injection of radiopaque microspheres (15 micron diameter) was performed in the left ventricle. The excised hearts were sliced in thin slices to measure the flow distribution up to a very small mass of tissue. With a technique called autoradiography the microspheres present in each slice were visualised as dark spots. Once digitised, the data were analysed to estimate the relative dispersion RD(m) of the mycrospheres for different sample weights.

5. Results

DFA (normal subject). Figure 3 shows the log-log plot of $F(n)$ versus n for a normal subject; a linear relationship is visible, with a slope $\alpha 2 = 1.02$, for fluctuations greater than about 10 beats. Below 10 beats a different slope $\alpha = 1.54$ can be detected.

DFA (transplanted subject). The transplanted subject also shows (Fig. 4) a linear relationship between $F(n)$ and n, with a clear cross-over approximately located at about 100 beats; the slope for small fluctuations (below 100 beats) has a value $\alpha 1 = 0.5$, while above 100 beats $\alpha 2 = 1.52$.

Blood flow distribution (canine heart). Figure 5 shows the log-log plot of RD(m) against the mass in grams. The linear relation is clearly visible: the slope of the regression line is s = -0.39 corresponding to a fractal dimension D = 1.39.

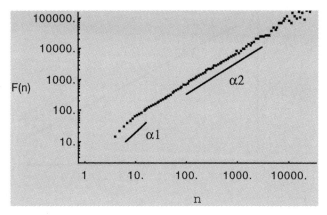

Figure 3. DFA of a normal subject. A slope of $\alpha 1 = 1.54$ is calculated for fluctuations below 10 beats; for fluctuations above this threshold the slope becomes $\alpha 2 = 1.02$.

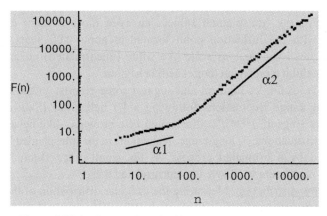

Figure 4. DFA of a transplanted subject. A cross-over behaviour is clearly visible around 100 beats, with a slope of $\alpha 1 = 0.50$ (below 100 beats) and $\alpha 2 = 1.52$ (above 100 beats).

6. Conclusions

Scaling properties. The DFA methods clearly indicated the existence of a scaling behaviour in the 24-hour HRV of both normal and transplanted subjects, even though important distinctions were revealed by the values of the scaling exponents. In the normal subject a behaviour suggestive of an underlying 1/f power-law process is detectable for fluctuations involving more than 10 beats; while a brownian-like behaviour could be hypothesised for small range fluctuations (below

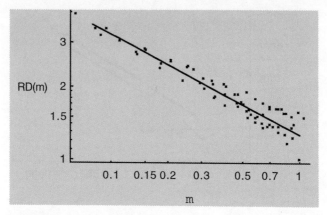

Figure 5. Relative dispersion of blood flow versus mass of tissue in double-log plot. The regression line has a slope of -0.39, corresponding to a fractal dimension D=1.39.

10 beats). For the transplanted subject an even more evident cross-over is detectable, with the inflection point located at about 100 beats. Below this "threshold" small fluctuations scale like white (uncorrelated) noise while large fluctuations exhibit a brownian (correlated) behaviour.

The above examples confirm and suggest some results: 1) the normal heart exhibits long range fluctuations following a 1/f behaviour (Saul *et al.*, 1988; Yamamoto & Hugson, 1994); 2) there could be a brownian-like behaviour for the small-range fluctuations; 3) long-range fluctuations in the transplanted heart seem to be generated by a brownian process; 4) the small-range fluctuations of the transplanted heart seem to behave like uncorrelated noise.

<u>Blood flow distribution</u>. Measuring the irregular distribution of the blood flow into the heart by means of the simple formula RD= SD/mean can be misleading, due to the "fractal" nature of this distribution: that makes the standard deviation to vary considerably with the size of tissue analysed, at least up to pieces of tissue as small as 2 mg of mass, as indicated by the overall results of the previously cited study.

References

Balocchi, R. *et al.* (1994) "Fractal Analysis of 24-hour Heart Rate Variability", in: *Intelligent Engineering Systems Through Artificial Neural Networks*, CH Dagli, BR Fernandez, J Ghosh, RT Soundar-Kumara, eds, New York, ASME Press, pp. 713-723.

Bassinghwaithe, J.B., R.B. King, S.A. Roger (1989) "Fractal Nature of regional myocardial blood flow heterogeneity", *Circ Res* 65: 578-590.

Di Garbo, A. *et al.* (1996) "Search for Nonlinearity in the Heartbeat Time Series", *Cybernetics and Systems*, 1: 575-579.

Giaquinta, A., S. Boccaletti, T. Arecchi (1996) "Superexcitability Induced Spiral Breakup in Excitable Systems", *Int. J. Bifurcation and Chaos* 9: 1753-1759.

Goldberger, AL. *et al.*, (1990) "Chaos and Fractals in Human Physiology", *Sci Am* 262: 42-49.

Holden, A.V., M.J. Poole, J.V. Tucker (1996) "An Algorithmic Model of the Mammalian Heart: Propagation, Vulnerability, Re-Entry and Fibrillation", *Int. J. Bifurcation and Chaos* 9: 1623-1635.

Iversen, P.O. & G. Nicolaysen (1995) "Fractals describe blood flow heterogeneity within skeletal muscle and within myocardium", *Am. J. Physiol.* 268: H112-H116.

Kaplan, D. & L. Glass (1995) *Understanding Nonlinear Dynamics*, New York: Springer Verlag.

Pagani, M. *et al.* (1986) "Power Spectral Analysis of Heart Rate and Arterial Pressure Variabilities as a Marker of Sympatho-Vagal Interaction in Man and Conscious Dog", *Circ Res* 59: 178-193.

Peng, C-K *et al.* (1995) "Quantification of Scaling Exponents and Crossover Phenomena in Nonstationary Heartbeat Time Series", *CHAOS* 5: 82-87.

Sugihara, G. & R. May (1990) "Nonlinear forecasting as a way of distinguishing chaos from measurement error in a data series", *Nature* Lond. 344: 734-741

Trivella, M.G., *et al.* (1996) "Microvascular perfusion unit visualized by computerized autoradiography of canine myocardium", *Proc. 6th World Congress for Microcirculation*, Munich (D), pp. 851-855.

NONLINEARITY OF NORMAL SINUS RHYTHM

MICHELE BARBI, SANTI CHILLEMI, ANGELO DI GARBO
Istituto di Biofisica del CNR, Via S. Lorenzo 26,56127-Pisa (Italy)
e-mail: barbi@ib.pi.cnr.it

and

RITA BALOCCHI, CLARA CARPEGGIANI, MICHELE EMDIN,
CLAUDIO MICHELASSI
Istituto di Fisiologia Clinica del CNR, Via Trieste 41, 56126-Pisa, Italy

ABSTRACT

Trying to answer the question whether the irregular heart activity is due to a chaotic process, short term predictability and nonlinearity were estimated on sequences of interbeat intervals from 24 subjects in relaxed conditions. The classical non-linear prediction method as well as a direct one, able to estimate the time series nonlinearity (Sugihara, 1994) were employed. The two approaches yield very similar results showing that a clear nonlinear behaviour is present in most of the examined sequences. Furthermore, the statistic produced by a completely different method as *detrended fluctuation analysis* (Peng et al., 1995) exhibits a strong correlation with the estimators of nonlinearity.

1. Introduction

There is a growing interest to characterizing the nature, stochastic or low-dimensional deterministic, of the irregular activities generated from both heart (Lefebvre *et al.*, 1993; Sugihara, 1994) and neuronal ensembles (Chang *et al.*, 1994; Schiff *et al.*, 1994). In particular, the hypothesis that cardiac rhythm may exhibit the features of deterministic chaos (Goldberger *et al.*, 1990) is still under debate (Kanters *et al.*, 1996; Vibe & Vesin, 1996). Thus, as a chaotic system is characterised by the short term predictability and the nonlinearity, it is worth investigating these features in the normal sinus rhythm.

Recently, a number of techniques for making predictions have been developed. Most of them aim to describe the short-term structure of the experimental time series by curves plotting either the normalised prediction error (Farmer & Sidorovich, 1987; Sauer, 1994) or the correlation coefficient of predicted to real values (Sugihara & May, 1990; Hoffman *et al.*, 1995) against the prediction time. These plots are useful to identify extreme behaviours (purely deterministic signal and white noise) but are not sufficient to reach clear conclusions in more complicated situations with unknown amounts of autocorrelated noise. Nevertheless this approach becomes more robust by carrying out suitable statistical controls on the temporal series under investigation. As the hallmark of a chaotic system is having a deterministic structure beyond the one corresponding to its autocorrelation, bootstrapping methods based on the comparison with *surrogate data* are mandatory (Theiler *et al.*, 1992). Usually, from each experimental series several surrogate ones, sharing some features of the original series (mean, standard deviation, autocorrelation function) but otherwise random, are derived. Then the prediction errors of both the experimental data and the surrogates are computed. As soon as the values obtained on the surrogates are statistically different from those on the data, the null hypothesis that the latter are generated by a linear stochastic process can be rejected.

To estimate the nonlinearity present in the experimental data it is also possible to employ the recently proposed *S-maps* method (Sugihara, 1994). Here, as an adimensional parameter is increased, the whole spectrum of models of the analysed system, from global linear to progressively more non-linear, is spanned. The corresponding predictability improvement estimates the system nonlinearity.

Both these techniques as well as *detrended fluctuation analysis* (Peng *et al.*, 1995) will be applied in the present paper to interbeat interval sequences obtained from healthy subjects in relaxation. An outline of the methods will be given in the next section. More information can be found in the full paper (Barbi *et al.*, 1997).

Also, due to the practical relevance of establishing whether the different discriminating statistics are independent measures or instead describe the same aspects of the dynamics, the correlations between one statistic and another will be inspected.

2. Materials and Methods

2.1. Data Collection

Continuous ECG signals were recorded from 24 healthy volunteers, during a 50 minutes session in supine position. Each subject was instructed, at the beginning of the session, to lie quiet and relaxed without speaking or moving.

The ECGs were sampled off line at 250 Hz. A suitable code allowed the successive R waves to be detected and their times of occurrence to be measured. For each subject, a rather stationary segment consisting of 1000 RR intervals, free of ectopic beats and erratic transients, was selected for the analysis. Also, to eliminate possible drifts of the mean beating rate, the sequence of interbeat intervals was preliminarily replaced by the sequence of their first differences.

2.2. Nonlinear Prediction and Surrogate Data Method

Let us outline now the first approach used for the analysis of selected time series, i.e. classical nonlinear prediction strengthened by the comparison with suitable surrogates.

Let $X_1, X_2, \ldots, X_n$ be the series of the first differenced RR intervals and m the embedding dimension such that the *delay vectors* $Xj = (X_j, \ldots, X_{j-(m-1)\tau})$ in $\mathbb{R}^m$ allow the trajectory to be disentangled. As many *predictees* X_i as possible are selected in the reconstructed space (Sugihara & May, 1990). For each predictee, a given number (usually $m+1$) of its nearest neighbours not sharing any coordinate with it are followed in their time evolution and the weighted average of their translations after k time steps is computed to get the corresponding prediction $X_{i,k}$. The *out-of-sample predictability*[1] of the time series $\varepsilon(k)$ is quantified by the adimensional ratio between the average prediction error and the series standard deviation σ (which can be regarded as the average prediction error obtained by assuming the mean of the time series as the same prediction for all times). Explicitly:

$$\varepsilon(k) = \frac{1}{\sigma}\sqrt{\frac{\sum_i (X_{i,k} - X_{i+k})^2}{N}} \tag{1}$$

where N is the number of predictees. This statistic will be referred to in the following as the *normalised prediction error* (NPE).

In the analysis reported here, a unity lag time and an embedding dimension $m = 7$ were adopted, respectively on inspection of the decay of the autocorrelation function and on the basis of preliminary calculations of NPE against m. This is also in keeping with other author's assumptions (Lefebvre *et al.*, 1994; Hoffman *et al.*, 1995).

Now, the mere observation of a short term predictability in the data is not sufficient to assess the low-dimensionality of the underlying system; coloured noise containing significant autocorrelation generates the same behaviour. As anticipated in the introduction, to exclude that the data predictability can be accounted for by a linear stochastic process, statistical controls based on surrogate data must be carried out. The *Fourier shuffled* surrogates used here have the same mean, standard deviation, and amplitude histogram as the original sequence, and a slightly changed autocorrelation (Theiler *et al.*, 1992, Chang *et al.*, 1994).

Once the surrogate sequences are created, their prediction error is computed and compared with that of the original series: a statistically meaningful difference invalidates the *null hypothesis* that the data predictability is accounted for by linear correlation in time.

Figure 1b plots, for some typical model sequences, the average value of the surrogate NPEs at one step into the future against the value of NPE computed on the original sequence. Bars quantifying the standard deviation σ_ε of the surrogate NPEs are also reported. It is worth noting that points 1 and 2 in Fig. 1b correspond

---

[1] We should better speak about *unpredictability*, as higher ε-values mean worse predictions.

to the same underlying system (the Henon map) but for point 2 some noise has been added. So noise makes the representative point to approach the identity line.

2.3. S-maps Method

For each *predictee* $X_i = (X_i, ..., X_{i-m+1})$ along the trajectory, let the prediction for the interval X_{i+k} (*k* steps into the future) be again $X_{i,k}$. One then writes (Longtin,1993; Sugihara, 1994):

$$X_{i,k} = C_0(i) + \sum_{j=1}^{m} C_j(i) X_{i-j+1}, \qquad i = m, ..., n-k. \qquad (2)$$

The coefficients $C_j(i)$ are obtained by solving the system of equations:

$$w_{hi} X_{h+k} = w_{hi} [C_0(i) + \sum_{j=1}^{m} C_j(i) X_{h-j+1}], \quad h = m, ..., n-k, \ |i-h| \geq m, \quad (3)$$

where the weights w_{hi} have the expression:

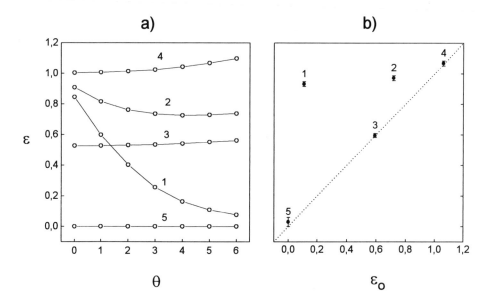

Fig. 1. Performance of the two analysis methods (see the text) on five typical sequences: 1) Henon map (chaotic regime); 2) Henon map plus white noise of 50% s.d.; 3) autocorrelated noise; 4) white noise; 5) deterministic (sinusoidal) signal. In (a) $\varepsilon(\theta,1)$ is plotted against θ, in (b) ε_S is plotted against ε_0.

$$w_{hi} = \exp(-\theta\, d_{hi} / <d_i>)\,, \qquad \theta \geq 0,\quad d_{hi} = |X_h - X_i|,$$

$<d_i>$ being the average distance of predictee from all other vectors. The restriction on the range of values spanned by index h guarantees an *out-of-sample* prediction.

System (3) is overdetermined so it must be solved by using either the direct least squares method or, as we did, the *singular value decomposition*. This approach was proposed by Sugihara (1994), who named it the *S-maps*. It can be considered a prediction method, which is linear for $\theta = 0$ (yielding $w = 1$ identically and C-coefficients approximately constant) but becomes the more non-linear the larger θ is made.

The series predictability is again quantified by the NPE as given in Eq. (1), but in this case it is dependent also on θ and so will be called $\varepsilon(\theta,k)$. As θ is increased, the map fitted to the system flow becomes more local and hence nonlinear; the corresponding decrease of $\varepsilon(\theta, k)$ (for a fixed prediction time k) directly estimates the nonlinearity of the time series (Sugihara, 1994).

For the sake of demonstration, Fig. 1a plots $\varepsilon(\theta,1)$ against θ for the same sequences of Fig. 1b. The initial slope of the curves in Fig. 1a and the distance of the corresponding points from the identity line in Fig. 1b appear to be clearly correlated.

In this paper the difference

$$\delta = \varepsilon(0, 1) - \varepsilon(1, 1) \qquad (4)$$

will be adopted as the representative statistic for this method.

2.4. Detrended Fluctuation Analysis (DFA)

This simple test has been proposed to quantify the long-range correlation behaviour of the sequences of heartbeat intervals (Peng *et al.*, 1995) and also of DNA bases (Peng *et al.*, 1994). Firstly, an integrated time series is derived from the original sequence of intervals. After dividing the corresponding *landscape* into n/l adjacent windows of size l, a straight line is fit to it in each window and the overall root-mean-square deviation $F(l)$ of the experimental values from the fitted lines is calculated. The procedure is started for $l = 4$ and then iterated for increasing l values.

Whenever $log\ F(l)$ scales with $log(l)$, or $F(l) \propto l^\alpha$, long term correlations exist in the sequence of intervals. This is what happens in the sequences analysed, where instead the crossover behaviour reported elsewhere (Peng *et al.*, 1995) is scarce or absent (data not shown). Therefore we simply calculated the *scaling coefficient* α by fitting a straight line to the plot of $log\ F(l)$ against l over the range (4-96). As those authors recommend to use rather long time series (on the order of 8000 intervals for their data), DFA was applied to the whole recording of each subject (of four to five thousand intervals). Moreover, the quantity $\beta = (\alpha - 0.5)^{-1}$ was adopted

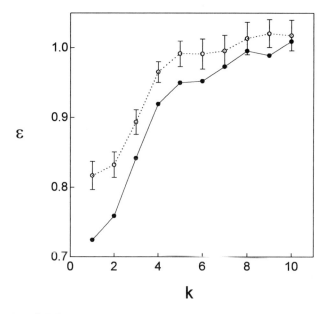

Fig. 2. Plot of ε against prediction time k for subject Cas. Solid points: original series; open points: surrogate series.

as the representative statistic. Note also that 0.5 is the α-value of the white noise (but also of the tent map) and separates the range of values corresponding to a positive correlation ($\alpha > 0.5$) from those implying a negative correlation ($\alpha < 0.5$).

3. Experimental Results

Let us start with the results obtained with nonlinear prediction and surrogate data.

For a typical case (subject Cas), NPE of the original series and average NPE of the surrogate series are plotted against prediction time k in Fig. 2. As k is increased, both NPEs gradually rise to an asymptotic value close to 1. A clear short term predictability is exhibited from the original series but also, though to a lesser extent, from the surrogate data. That could be anticipated by simply inspecting the time series autocorrelation function (Fig. 3).

As the most information about non-linear predictability is carried out by $\varepsilon(1)$, and considering the high number of subjects to analyse, we decided to limit our attention to the first point of the predictability curves. This allowed us also to simplify the notations by omitting in them the information about prediction time.

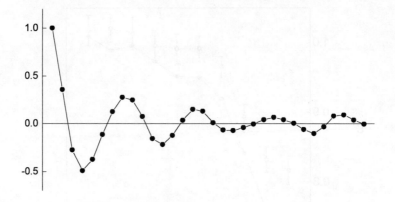

Fig. 3. Autocorrelation function of subject Cas sequence.

For each of the 24 subjects, ten Fourier shuffled surrogate time series were generated from the original one, and $\varepsilon(1)$ was calculated for both the original and the surrogate sequences. The results are shown in Fig. 4; therein, for each subject,

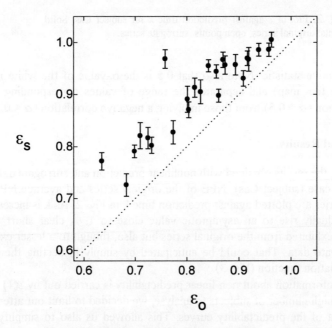

Fig. 4. The average value of NPE for surrogates is plotted against its value for the original series. Error bars as in Fig. 1.

ε_o is plotted against ε_s, while the bars of total length $2\sigma_\varepsilon$ delimit the 68% probability range around these averages. All the representative points fall above the identity line and, most of them, consistently distant from it (in units of σ_ε). The short distances of a few points from the identity line can be ascribed to a consistent amount of "noise" present in the sinus node pacemaker (see again Fig. 1b).

By using the *phase-randomised* surrogates (Theiler *et al.*, 1992, Chang *et al.*, 1994) similar results are obtained (data not shown). Summarising, the overall picture suggests the conclusion that a clear nonlinearity exists in the analysed time series.

Coming now to the S-maps method, the values of the δ-statistic were computed for the same 24 sequences and plotted against the corresponding values of the "nonlinear predictability" $\varepsilon_s - \varepsilon_o$ in Fig. 5. A strong correlation between the two quantities, confirming our assessment of the time series nonlinearity, is evident.

To further characterise these heart rhythms, the DFA was applied to them. For each of the 24 sequences, the scaling coefficient α was calculated, and the quantity $\beta = (\alpha - 0.5)^{-1}$ was plotted against the corresponding value of $\varepsilon_s - \varepsilon_o$ (Fig. 6). The two statistics appear to be strongly correlated ($\rho = 0.92$, corresponding to $p < 10^{-4}$). Also, it is worth noting that neither ε_s nor ε_o correlate with β.

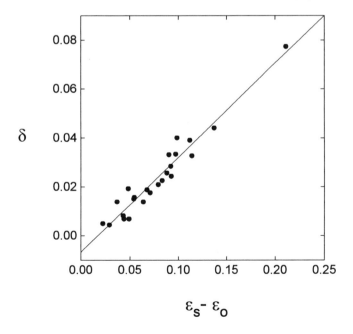

Fig. 5. Correlation between the two methods ($\rho = 0.97$).

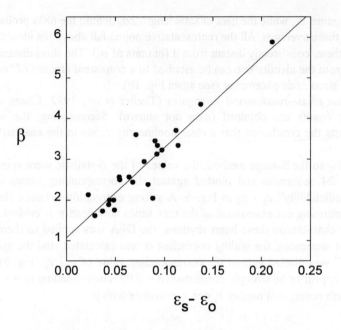

Fig. 6. Plot of the β statistic of DFA against $\varepsilon_s - \varepsilon_0$.

4. Discussion

The results obtained on a sample of 24 healthy subjects in relaxed conditions, by the two approaches of the S-maps and the nonlinear predictors, confirm the presence of nonlinearity in the time series analysed, giving similar estimates and validating each other. So the question whether nonlinearities reflect some physiological information will be worthy of investigation by comparing the results obtained on groups of subjects in different physiological conditions.

While the excellent agreement between the results of these two methods could be expected, as both are designed to just reveal an underlying nonlinear dynamics, the strict correlation of the nonlinear predictability $\varepsilon_s - \varepsilon_0$ with the β statistic adopted for DFA is a truly new finding. In fact, this last method was devised to simply characterise the tail of the time series autocorrelation. However, the correlation between the two statistics does not mean that β statistic can be used in general as a marker of nonlinearity - it is enough to observe that α values between 0.5 and 1 can be obtained from colored noise too. What Fig. 6 implies is that, for the RR interval series analysed here, β and $\varepsilon_s - \varepsilon_0$ are not independent measures but describe the same aspects of the dynamics. In particular, the series with stronger nonlinearity are characterised by autocorrelation functions which decay very fastly (a requisite of low-dimensional chaos).

Lastly, the non-linear predictability exhibited by most subjects examined implies that chaotic behaviour is not ruled out, but deeper analysis is needed to reach a definite conclusion. Also, nonlinear predictability should be further investigated to test its dependence on the modulation of sinus node activity by physiological stimuli like respiration, baroreflex control, termoregulation, etc.

References

Barbi, M., S.Chillemi, A. Di Garbo, R. Balocchi, C. Carpeggiani, M. Emdin, C. Michelassi & E. Santarcangelo (1997) "Predictability and nonlinearity of the heart rhythm", *Chaos, Solitons and Fractals*. In press.

Chang, T., S.J. Schiff, T. Sauer, J.P. Gossard & R.E. Burke (1994) "Stochastic Versus Deterministic Variability in Simple Neuronal Circuits: I. Monosynaptic Spinal Cord Reflexes", *Biophys. J.* 67: 671-683.

Farmer, J.D. & J.J. Sidorowich (1987) "Predicting chaotic series", *Phys. Rev. Lett.* 59: 845-848.

Goldberger, A.L., D.R. Rigney & B.J. West (1990) "Chaos and fractal in human physiology", *Sci. Am.* 262: 42-49.

Hoffman R.E., W. Shi & B.S. Bunney (1995) "Nonlinear Sequence-Dependent Structure of Nigral Dopamine Neuron Interspike Interval Firing Patterns", *Biophys. J.* 69: 128-137.

Kanters, J.K., M.V. Højgaard, E. Agner & N. Holstein-Rathlou (1996) "Short and long-term variations in non-linear dynamics of heart rate variability", *Cardiovascular Research* 31: 400-409.

Kaplan, D.T., M.I. Furman, S.M. Pincus, S.M. Ryan, L.A. Lipsitz & A.L. Goldberger (1991) "Aging and the complexity of cardiovascular dynamics", *Biophys. J.* 59: 945-949.

Kaplan, D.T. & L. Glass (1995) "Understanding nonlinear dynamics", Springer Verlag.

Kleiger, R.E., J.P.Miller, J.T. Bigger & A.J. Moss (1987) "Decreased heart rate variability and its association with increased mortality after acute myocardial infarction", *Am. J. Cardiol.* 59: 256-262.

Lefebvre, J.H., D.A. Goodings, M.V. Kamoth & E.L. Fallen (1993) "Predictability of normal heart rhythms and deterministic chaos", *Chaos* 3: 267-276.

Longtin, A. (1993) "Nonlinear forecasting of spike trains from sensory neurons", *Int. J. Bif. and Chaos* 3: 651-661.

Peng, C.K., S. Havlin, H.E. Stanley & A.L. Goldberger (1995) "Quantification of scaling exponents and crossover phenomena in nonstationary heartbeat time series", *Chaos* 5: 82-87.

Peng, C.K., S.V. Buldyrev, S. Havlin, M. Simons, H.E. Stanley & A.L. Goldberger (1994) "Mosaic organization of DNA nucleotides", *Phys. Rev. E* 49: 1685-1689.
Sauer, T. (1994) "Reconstruction of dynamical systems from interspike intervals", *Phys. Rev. Lett.* 72: 3811-3814.
Schiff, S.J., K. Jerger, T. Chang, T. Sauer & P.G. Aitken (1994) "Stochastic versus deterministic variability in simple neuronal circuits: II Hippocampal slice", *Biophys. J.* 67: 684-691.
Sugihara, G. (1994) "Nonlinear forecasting for the classification of natural time series", *Phil. Trans. R. Soc.* A 348: 477-495.
Sugihara, G. & R. May (1990) "Nonlinear forecasting as a way of distinguishing chaos from measurement error in a data series", *Nature* 344: 734-741.
Theiler, J., B. Galdrikian, A. Longtin, A. Eubank & J.D. Farmer (1992) "Testing for nonlinearity in time series: the method of surrogate data", *Physica* D 58: 77-94.
Vibe, K. & J. Vesin (1996) "On chaos detection methods", *Int. J. Bif. and Chaos* 6: 529-543.

CHAOS CONTROL AND ITS APPLICATIONS

ADAPTIVE STRATEGIES FOR RECOGNITION AND CONTROL OF CHAOS

F.T. ARECCHI[(+)], S. BOCCALETTI
Istituto Nazionale di Ottica, Largo E. Fermi, 6,
50125 Firenze (Italy)
e-mail arecchi@fox.ino.it
(+) Also Dept. of Physics, University of Firenze,
50125 Firenze (Italy)

ABSTRACT

Combining knowledge of the local variation rates as well as of the long time trends of a dynamical system, we introduce an adaptive recognition technique consisting in a sequence of variable resolution observation intervals at which the geometrical positions are sampled. The sampling times are chosen so that the sequence of observed points forms a regularized set, in the sense that the separation of adjacent points is almost uniform. As a result, information on the dynamical features is contained in the non regular sequence of sampling times. Extracting statistical information from this one-dimensional sequence is easier than spanning the N-dimensional coordinate space. This adaptive technique is able to recognize the unstable periodic orbits embedded within a chaotic attractor and stabilize anyone of them even in the presence of noise, through small additive corrections to the dynamics.

1. Introduction

Standard investigation of a dynamical system is carried out by fixing a sequence of regularly spaced observation times and plotting the corresponding positions in some coordinate space.

The geometry of such a time series of data discriminates on whether a signal is regular, chaotic, or random (Grassberger and Procaccia, 1983; Abarbanel *et al.* 1993). An alternative strategy introduced by us (Arecchi *et al.*, 1994) consists in fixing a given resolution in the coordinate space and registering only the subset of the whole signal consisting of those points such that the separation of two successive ones is within that resolution. The corresponding geometric set is an almost regular one, but the sequence of stroboscopic observation times is now erratic when the considered dynamics is irregular. At variance with the geometry of a time series, which yields a set of points embedded in a N dimensional space, the stroboscopic times provide a one dimensional string of data.

This implies that for such a strategy the indicators of chaotic motion, based on the mutual distances of points in state space (Grassberger and Procaccia, 1983) are not manageable. On the contrary, all those indicators relying on the time information, as e.g. the maximum Liapunov exponent, the location and stability of the unstable periodic orbits of a chaotic attractor and the signal to noise ratio within

a frequency range, are easily accessible to our method with a computation time which increases linearly rather than exponentially with the number of dimensions. As a result, application of these time indicators has become more convenient as we will show.
In this lecture we review the results of recent studies, whereby this adaptive strategy introduced for recognition (Sec. 2) is applied to control (Sec. 3).

2. Chaos Recognition

The standard description of a dynamical system is done by selecting a uniform sequence of observation times (i.e., separated by a constant time interval) and plotting the evolution of the corresponding positions in some coordinate space. Based upon the geometry of the time series, one extracts suitable indicators to classify a signal as regular, chaotic, or random (Abarbanel et al., 1993).

Whenever the motion is confined within a finite region of space, an alternative approach consists in fixing a narrow observation window in the coordinate space. Registering only data within the window is a kind of stroboscopic inspection that provides a clustered set of geometric positions, but now the sequence of return times to the window is erratic if the dynamics is irregular. An adaptive windowing overcomes the above difficultiy (Arecchi et al., 1994). Precisely, we position the sampling times upon the information provided by the local variation (expansion or contraction) rates, probed with a sensitivity which is readjusted to cope with the long time trends of the dynamics. The stroboscopic time sequence that we extract provides not only useful indicators for the chaotic case, but it also permits discrimination between deterministic and stochastic dynamics. Let us consider a dissipative dynamics $\dot{x} = f(x,\mu)$, where x is a D-dimensional vector and μ a set of control parameters, and start by setting μ in order to have a stable periodic orbit of period $\tilde{\tau}$. A $(D-1)$ - dimensional Poincaré section intercepts the orbit at a point which repeats after a time $\tilde{\tau}$. Suppose next that, by a change of μ, the orbit is destabilized toward a chaotic attractor.

Rather than modifying the trajectory as in control methods, we aim at recognizing its unperturbed features. For this purpose, based upon the information provided by the local variation rates, we make the next observation interval shorter or longer than $\tilde{\tau}$, in order to minimize the variation in width of the window which includes the two points at the extreme of the interval. This depends crucially on the fact that the expansion or contraction in a given direction i cannot be monotonic in the course of time, since the attractor is confined within a finite support and hence the trajectory undergoes frequent twistings.

Precisely, for each coordinate axis i (i=1....,D), we consider the variation

$$\delta x_i(t_{n+1}) = x_i(t_{n+1}) - x_i(t_n), \tag{1}$$

where $t_{n=1} - t_n = \tau_n$ is the nth adjustable interval, to be specified. The goal of the adaptive windowing is to keep the separation between two successively observed coordinates as stable as possible, that is, to minimize the variations of δx_i. In order to assign τ_{n+1} we consider the local variation rate

$$\lambda_i(t_{n+1}) = \frac{1}{\tau_n} \log \left| \frac{\delta x_i(t_{n+1})}{\delta x_i(t_n)} \right|. \tag{2}$$

Here τ_n is the minimum of all $\tau_n^{(i)}$ corresponding to all different i, updated by the rule

$$\tau_{n+1}^{(i)} = \tau_n^{(i)} \left(1 - \tanh(\sigma \lambda_i(t_{n+1}))\right). \tag{3}$$

The hyperbolic tangent function maps the whole range of $\sigma \lambda_i$ into the interval (-1,+1). The constant σ, strictly positive, is chosen in such a way as to forbid $\tau_{n+1}^{(i)}$ from going to zero. It may be taken as an apriori sensitivity. A more sensible strategy would consist in looking at the unbiased dynamical evolution for a while and then taking a σ value smaller than the reciprocal of the maximal λ recorded in that time span. Fixing σ is like fixing the connectivities of a neural network by a preliminary learning session, while adjusting σ upon the information accumulated over previous time steps corresponds to considering σ as a kind of long-term-memory, adding some extra correction to the short-term-memory represented by the sequence of τ_n.

For a fixed σ, the steps of the algorithm are: observation of the data at t_{n+1}; evaluation of $\delta x_i(t_{n+1})$; evaluation of $\lambda x_i(t_{n+1})$; updating of $\tau_{n+1}^{(i)}$; selection of $\min_i \tau_{n+1}^{(i)} = \tau_{n+1}$; determination of the new strobing time $t_{n+2} = t_{n+1} + \tau_{n+1}$. This way, we obtain a sequence of observation times starting from given t_0 and $\tilde{\tau}$

$$t_0, t_1 = t_0 + \tilde{\tau}, t_2 = t_1 + \tau_1, \ldots t_{n+1} = t_n + \tau_n, \ldots \tag{4}$$

corresponding to which the variations of $\delta x_i(t_n)$ can be reduced below a preassigned value. The positions observed at the times t_n are confined within the adaptive window enfolding the trajectory; however the time sequence (Lorenz, 1963) now includes the chaotic information which was in the original geometric sequence x(t). Combination of the successive relocation of each τ value with the long time readjustment of the sensitivity σ provides this adaptive strategy with two separate hierarchical levels of control, and hence it is able to cope with singular situations as shown later. In Ref. (Arecchi,*et al.*, 1994) we have applied the method to many chaotic situations as the Lorenz model (Lo) (Lorenz, 1963), the 3 and 4 dimensional Rossler model (called respectively Ro3 (Rossler, 1976) and Ro4 (Rossler, 1979)) and the Mackey-Glass delay equation (MG) (Mackey and Glass, 1977).

The sequence of strobing intervals contains the relevant information on the dynamics, thus we can characterize chaos as follows. Since in Eq. (3) $|\sigma\lambda_i| \ll 1$, then two successive τ_n must be strongly correlated. Thus, even though the τ_n may be spread over a rather wide support, the return map τ_{n+1} vs. τ_{n+1} vs. τ_n clusters along the diagonal. Any appreciable deviation from the diagonal denotes the presence of uncorrelated noise. This is shown in Fig.1 where we plot the return map of the τ_n for Ro4 and for Ro4 with 1% noise.

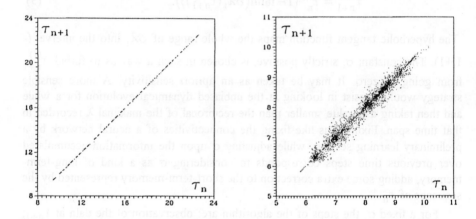

Fig. 1. Return maps τ_{n+1} vs. τ_n for a) Ro4 and b) Ro4 with an additional 1% white noise.

To extract a quantitative indicator from the return map τ_{n+1} vs. τ_n, we evaluate the average deviation of the observation times away from the diagonal. This should provide an estimate for the maximum Liapunov exponent Λ_{Max} since, by Eq. (3), the normalized deviation yields the local rate. By direct use of Eq. (2), we propose the following indicator

$$\eta = \frac{1}{M} \sum_+ \sum_i \lambda_i(t_n), \qquad (5)$$

where $\sum_+$ accounts only for the n_+ cases for which min $\tau_{n+1}^{(i)} \leq \min \tau_{n+1}^{(i)}$, $\sum_i$ runs over all m dimensions of the phase space, and M=Nm, where $N = n_+ + n_-$ is the total number of strobing observations ($n_{+(-)}$ is the number of the shrinking (stretching) strobing intervals). The soundness of Eq. (5) is shown by comparing η with Λ_{Max} in the case of Ro3. The positive Liapunov exponent of Ro3 with a=b=0.2, c=5.7 and for N_d=40,000 data turns out to be Λ_{Max}=0.072± 0.001\$ by the Benettin et al. algorithm (Benettin et al., 1980) and Λ_{Max}=0.075± 0.005 by the Sano-Sawada algorithm (Sano and Sawada, 1985). Applying our method with a sensitivity g=0.00038 to the same system ($\tilde{\tau}$=6.3) and evaluating η for an increasing number of data, η reaches a plateau beyond 6,000 data. Its average between 8,000 and 12,000 data, with three times the standard deviation is η=0.0722 ±0.0003.

So far we have dealt with an assigned model as Ro3, Ro4 and Lo. A more difficult task is to characterize an empirical time series, in order to discriminate whether it corresponds to a deterministic or stochastic phenomenon. For this purpose we embed our data series in spaces of increasing dimensions. A very stringent test is represented by MG

$$\dot{x} = -0.1 x(t) + \frac{0.2 x(t-\tau)}{1 + x(t-\tau)^{10}}. \quad (6)$$

With τ=100 it corresponds to a ~7.5 dimensional chaotic dynamics.

Let us consider the time series for MG, for a pure white noise with r.m.s. fluctuations as MG, and for a random phase time series having the same spectral power as MG (surrogate) (Theiler et al., 1992).

For an empirical series of data with no apriori information, a discrimination between determinism and noise is provided by the following indicator

$$\beta = \frac{1}{N} \sum_n \left| \prod_{i=1}^{m} \lambda_i(t_n) \right| \quad (7)$$

Its heuristic meaning emerges from the following considerations. Expanding Eq. (2) to first order and referring to the unit time step $\tau_n = 1$, we can write $\lambda_i(t_{n+1}) = (\delta x_i(t_{n+1}) - \delta x_1(t_n))/\delta x_i(t_n)$. We now evaluate the variation over the unit time of the volume $V_n = \prod_{i=1}^{m} \delta x_i(t_n)$ made by all m measured variations at time t_n. The relative variation rate $r_n = (V_{n+1} - V_n)/V_n$ is given by

$$r_n = \sum_i \lambda_i + \sum_{i \neq j} \lambda_i \lambda_j + \ldots + \prod_i \lambda_i \quad (8)$$

Summing up over all directions of phase space, we introduce the directional averages

$$\langle \lambda \rangle = \frac{1}{m}\sum_i \lambda_i, \langle \lambda^2 \rangle = \frac{2}{m(m-1)}\sum_{i \neq j}\lambda_i\lambda_j, \text{etc.}$$

As we further sum over all n up to N, the twisting along the chaotic trajectory makes all directions statistically equivalent, thus in $\sum_n r_n$ we can replace $\langle \lambda^k \rangle$ by $\langle \lambda \rangle^k$ for $2 \leq k \leq m$. In the case of stochastic noise, since variations over successive time steps are uncorrelated, $\delta x(t_{n+1}) - \delta x(t_n) \cong \delta x(t_n)$ and $\langle \lambda \rangle$ is close to 1, so that $\langle \lambda \rangle^m = O(1)$. Instead for a deterministic dynamics two successive steps are strongly correlated, hence $\langle \lambda \rangle < 1$, and the last term of Eq. (8), that is $\langle \lambda^m \rangle \sim \langle \lambda \rangle^m = \exp(-m\log(1/<\lambda>))$ yields the most sensitive test.

Based on these considerations, we take the sum over the N trajectory points of the last term of Eq. (8) as the β indicator displayed in Eq. (7). In view of what said above, β scales as e^{-m} for a deterministic signal (asides a factor O(1) in the exponent) whereas it scales as e^0 for noise and as $\varepsilon^{-0.5m}$ for surrogate since one half of the total number of degrees of freedom (phases of the Fourier components) scale like noise and the other half (amplitudes) scale like the deterministic signal.

Fig. 2 shows the β plots vs. m for MG and its surrogate as well as for white noise. In order to estimate the minimal number S(m) of data necessary for any value m, we start from a very large value S and reduce it until we obtain a deviation $\Delta\beta/\beta \cong 10\%$. With this definition the number S(m) of data necessary for MG scales as S(m)=a+bm, with $a \cong 10,000$ and $b \cong 3,000$.

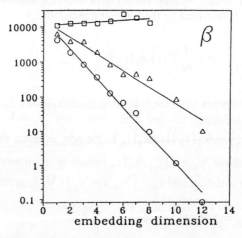

Fig. 2. β plots vs. embedding dimension m. Squares: white noise, triangles: surrogate of MG, circles: MG. For all cases σ = 0.0048. Solid lines are exponential fits $e^{(Am)}$ with A=-0.526 for the surrogate of MG and A = -0.987 for MG.

Comparing our adaptive recognition with statistical methods based upon the assignement of a probability measure in phase-space, as counting the number of neighbors within a given distance from each phase-point or the distribution of distances for closest neighbors, we easily realize that, for S data points, the number of computing operations scales as S in our case and as S^2 in the statistical cases. Furthermore, to assure a good resolution in m embedding dimensions, statistical methods require that S(m) increase exponentially in m, whereas our adaptive strategy is based on variation rates along the coordinate axes and hence our S(m) scales linearly with m as shown above.

3. Control of Chaos

Controlling chaos means stabilizing one of the unstable periodic orbit (UPO) (Auerbach *et al.*, 1987) of a chaotic dynamics by very tiny perturbations which do not affect the main dynamical features.

The different methods for controlling chaos can be classified into two main classes, namely: feedback methods and open loop methods. The first class includes the method proposed by Ott, Grebogi and Yorke (OGY) (Ott *et al.*, 1990), which consists in readjusting a control parameter each time the trajectory crosses the Poincare' section (PS), and that of Pyragas (Pyragas, 1995), based upon a continuous application of a delayed feedback upon one of the system variables.

The second class includes the method of Huebler (Plapp and Hübler, 1990; Jackson and Hübler, 1990), which presupposes knowledge of model equations and specifies a goal dynamics to construct control forces, as well as those methods which suppress chaos by means of periodic (Lima and Pettini, 1990) or stochastic (Braiman and Goldhirsch, 1991) perturbations to the system.

On the other hand, many experimental systems have been studied with the aim of establishing control over chaos, namely: a thermal convection loop (Singer *et al.*, 1991), a yttrium iron garnet oscillator (Azevedo and Rezende, 1991), a diode resonator (Hunt, 1991), an optical multimode chaotic solid-state laser (Roy *et al.*, 1992), a Belouzov-Zabotinsky reaction diffusion chemical system (Petrov *et al.*, 1993), a CO_2 single mode laser with modulation of losses (Meucci *et al.*, 1994).

Open loop or non-feedback methods are rather limited in their purposes, since they are applicable only to specific dynamical situations already described by a satisfactory model. Furthermore open loop perturbations are not able to select any of the UPO's, but they are limited to some specific dynamical ranges i.e. they are "goal oriented". On the contrary, eedback methods do not presuppose apriori knowledge of the model equations, and, after a sufficiently long inspection of the motion features, their control can be applied to any of the UPO's.

All feedback techniques (Ott *et al.*, 1995) are affected by some drawback. OGY needs a long acquisition section to retrieve the stable and unstable manifolds of the PS saddle point to be controlled. On the other hand, Pyragas controls the UPO's of a system perturbed by a delayed term, hence it is intrinsically based on a delay-differential dynamics which introduces any more dimensions than they were in the original one, including possible spurious UPO's.

Our adaptive technique provides a natural implementation of a feedback control which overcomes the drawbacks sketched above (Boccaletti and Arecchi, 1995), for the following reasons that we will specify later below: i) we do not need to retrieve the stability properties of the saddle points, hence the acquisition time is drastically reduced; ii) in our method we first measure the UPO's periods of the unperturbed dynamics, and then apply a control to stabilize a selected UPO; in this operation no spurious UPO's can appear. Let us consider the general dynamical dissipative system $\dot{x} = f(x, \mu)$ where x is a D-dimensional vector, f a nonlinear function and the control parameters µ are choosen as to produce chaos. For sake of exemplification, we refer to a case with more than one positive Liapunov exponent, precisely to the 4-dimensional Rossler (Ro4) model, described by the equations

$$\dot{x}_1 = -x_2 - x_3$$
$$\dot{x}_2 = x_1 + 0.25x_2 + x_4$$
$$\dot{x}_3 = 3 + x_1 x_3$$
$$\dot{x}_4 = -0.5x_3 + 0.05x_4$$

(9)

The initial conditions $x_1(0) = -20, x_2(0) = x_3(0) = 0, x_4(0) = 15$, generates a dynamics with two positive Liapunov exponents (Rossler, 1979).

The first step of the controlling strategy is to extract the periods of UPO's embedded within the chaotic attractor, by exploiting the sequence of observation time intervals produced as in Sec. 2. We then consider the maps τ_{n+k} vs. $\tau_n, k = 1, 2, ...$ and plot the r.m.s. η of the point distribution around the diagonal of such maps as a function of k. For a chaotic dynamics, temporal correlations decay over a finite time. Hence, one should expect $\eta(k)$ to be a monotonically increasing function. In fact, the dynamics brings the trajectory to shadow neighborhoods of different UPO's. As it gets close to an UPO of period T_j, the correlation is rebuilt and the distribution of τ includes windows of correlated values appearing as minima of η vs. k around $k_j = T_j / <\tau>, <\tau>$ being the average of the τ distribution.

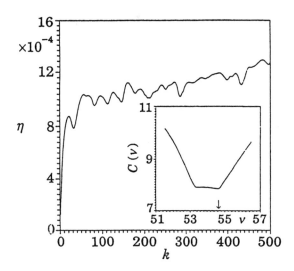

Fig. 3. η - k plot for Ro4. Initial conditions and parameters in the text. σ=0.01. Inset: cost function C(υ) for the eighth k minimum of the η curve. ν = 54.64 (indicated by the arrow) is the measurement of the period for one of the orbits 8 of Ro4.

Fig. 3 reports the η-k plot for Ro4, from which one can extract the different UPO's periods as the minima of the η curve. However, since during the observation the time intervals are changing, a rigorous determination of the period can be provided in the following way.

Let τ_{min} and τ_{max} be respectively the minimum and the maximum τ during the observation. The period T_j of the jth UPO is such that $k_j \tau_{min} \leq T_j \leq k_j \tau_{max}$ where k_j is the jth minimum of the η curve. We introduce a cost function

$$C(\nu) = \sum_{n=1}^{N} |x(t_n) - x(t_n - \nu)| \qquad (10)$$

where the sum runs over the N data recorded, and we look for the minimum of C(υ) with υ ranging from $k_j \tau_{min}$ to $k_j \tau_{max}$. Such a minimum measures the period T_j of the jth UPO. The inset of Fig. 3 shows the cost function C(υ) for the eighth minimum of the η plot. ν = 54.64 is the period of the 8th UPO of Ro4.

Once periods T_j (j=1,2,...) of the UPO's have been extracted, the second step is to achieve stabilization of the desired UPO when the system naturally visits closely phase space neighborhoods of that UPO. At each new observation time $t_{n+1} = t_n + \tau_n$ and for each component i of the dynamics, instead of Eq. (1), we evaluate the differences between actual and desired values:

$$\delta x_i(t_{n+1}) = x_i(t_{n+1}) - x_i(t_{n+1} - T_j), \tag{11}$$

and the local variation rates λ's now are

$$\lambda_i(t_{n+1}) = \frac{1}{\tau_n} \log \left| \frac{x_i(t_{n+1}) - x_i(t_{n+1} - T_j)}{x_i(t_n) - x_i(t_n - T_j)} \right|. \tag{12}$$

Eq. (3) and choice of the minimum are kept for the updating process of τ's.

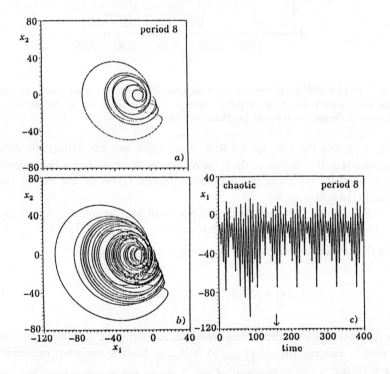

Fig. 4. (x_1, x_2) projection of the phase space portrait for (a) the controlled period 8 of Ro4, (b) the uncontrolled Ro4. $\sigma = 10^{-5}$. (c) Time evolution of the first component x_1 before and after control. Arrow indicates the instance at which control task begins.

We define U(t) as the vector with i^{th} component (constant over each adaptive time interval) given by

$$U_i(t_{n+1}) = \frac{1}{\tau_{n+1}}\left(x_i(t_{n+1} - T_j) - x_i(t_{n+1})\right), \quad (13)$$

and we add such a vector to the evolution equation, which now becomes

$$\frac{dx}{dt} = f(x,\mu) + U(t) \quad (14)$$

The λ's of Eq. (12) measure how the distance between actual and desired trajectory evolves. λ negative means that the true orbit is collapsing into the desired UPO, while λ positive implies that the actual trajectory is diverging away from the UPO and control has to be performed in order to constrain it to shadow the UPO. Contraction or expansion of τ's result in perturbing the dynamics more or less robustly to stabilize the desired UPO, by fixing the weight of the correction, which, once a given T_j has been imposed, is selected by the same adaptive dynamics. The introduced adaptive weighting procedure in Eq. (13) assures the effectiveness of the method (perturbation is larger or smaller whenever it has to be).

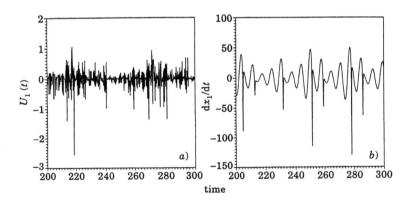

Fig. 5. (a) Temporal evolution of U_1 (see text for definition) and (b) corresponding evolution of dx_1/dt. Same stipulations for the control task as mentioned in the caption of fig. 4.

In Fig.4 we show the control of period-8 of Ro4. In Fig.5 we report $U_1(t)$ and $G_1(t)$ for the Ro4 model during control of period-8. It is evident that the former is between two and three orders of magnitude smaller than the latter as expected by the above discussion.

As for time scales, while in Eq. (11) differences δx_i are evaluated with respect to the goal dynamics (thus over the period T_j), in Eq. (12) all λ's are evaluated over the adaptive τ, which has to be larger than the Runge-Kutta integration interval (Arecchi et al., 1994) but smaller than the UPO's period. Thus the method introduces a natural adaptation time scale intermediate between minimum resolution time and time scale of the periodic orbits.

For practical purposes, we can approximate the denominator of Eq. (13) as

$$\tau_{n+1} = \tau_n\left(1 - \tanh(\sigma\lambda_{n+1})\right) \cong <\tau>\left(1 - \sigma\lambda_{n+1}\right) \tag{15}$$

where i) we have replaced τ_n with the ensemble average, and ii) we have linearized the tgh function. Now, λ_{n+1} of Eq. (12) can also be linearized as

$$\lambda(t) \cong \frac{1}{<\tau>}\frac{\dot{x}(t) - \dot{x}(t - T_j)}{x(t) - x(t - T_j)} \tag{16}$$

where we have approximated one discretized stroboscopic observation with a continuous inspection. Combining (15) and (16) into Eq. (13) this reduces to

$$U(t) = K_1\left(x(t - T_j) - x(t)\right) + K_2\left(\dot{x}(t - T_j) - \dot{x}(t)\right) \tag{17}$$

Where $K_1 = 1/<\tau>$ and $K_2 = \sigma/<\tau>^2$. This approximation has been suggested by R. Genesio. Its consequences are relevant.

The improvement over Pyragas ($K_2=0$) can be appreciated in a very heuristic way. One single adaptation parameter as K_1 might trap the system within a relative minimum, whereas the presence of two degrees of freedom (K_1 and K_2) allows a global optimization. Furthermore, recently a linearized perturbation technique in terms of a wash out filter has been introduced (Basso et al., 1997). It consists in dressing the feedback loop with a filter which has zeroes at two frequencies ($\omega = 0$ and $\omega = 1/T_j$). The filter has the following transfer function s=iω being the frequency domain)

$$W(s) = K_c \frac{s\left(s^2 + \Omega^2\right)}{\left(s^2 + \xi\Omega s + \Omega^2/4\right)(s + \nu\Omega)} \tag{18}$$

where K_c is a gain factor, and $T_j = 2\pi/\Omega$ is the period to be stabilized. For appropriate values of ξ and υ see Ref. 24.

Such a wash out filter is of course fully equivalent to Pyragas, in first approximation. Furthermore the high frequency features introduced by the delay operator $1 - e^{-i\omega T_j}$ corresponding to the Pyragas' method are put off by the polynomial approximation of the filter (18). This filter has been applied to the feedback stabilization of a CO_2 laser (Meucci et al., 1996). We notice that the linearized version of our adaptive technique reported in Eq. (17) is equivalent to introduce the new filter

$$(K_1 + sK_2)W(s) \qquad (19)$$

Eq. (19) allows a very convenient hardware implementation of our adaptive control. Since no computational time is prerequired, it is very appropriate for fast chaotic cases, such as laser dynamics.

An extended version of Pyragas technique has been introduced by Bleich and Socolar (Bleich and Socolar, 1996) and later elaborated by Pyragas (Pyragas, 1992). It consists in a correction signal which is a sum of contributions at all previous multiple pT_j (p integer) of the choosen period. The above criticism about the high dimensionality of the differential delayed dynamics holds even more. On the other hand, trying to increase the resolution by adding information at previous times is already the virtue of our adaptive technique, but in our case this is done over a set of close-by times, instead of a sparse sampling of the signal at time separations as long as T_j.

In a recent paper (Boccaletti et al., 1997), we have shown the effectiveness of our adaptive control method in a high dimensional dynamics, in the case of a delayed system.

References

Abarbanel, H.D.I., R. Brown, J.J. Sidorowich, L. Sh. Tsimring (1993) "The analysis of observed chaotic data in physical systems", *Rev. Mod. Phys.*, 65, 1331-1392;

Arecchi, F.T., G. Basti, S. Boccaletti, A.L. Perrone (1994) "Adaptive recognition of a chaotic dynamics", *Europhys. Lett.*, 26, 327-332.

Auerbach, D., P. Cvitanovic, J.P. Eckmann, G. Gunaratne, I. Procaccia (1987) "Exploring chaotic motion through periodic orbits", *Phys. Rev. Lett.*, 58, 2387-2389.

Azevedo, A. and M. Rezend, (1991), "Controlling chaos in spin-wave instabilities", *Phys. Rev. Lett.*, 66, 1342-1345.

Basso, M., R. Genesio, A. Tesi., (1997), "Controller design for extending periodic dynamics of a chaotic CO_2 laser", *Syst. and Contr. Lett.*, 31, 287-297.

Benettin, G., L. Galgani, A. Giorgilli, J.M. Strelcyn, (1980), "Lyapunov characteristic exponents for smooth dynamical systems and for Hamiltonian systems; a method for computing all of them", *Meccanica*, 15, 9-20.

Bleich, M.E., and J.E.S. Socolar, (1996), "Stability of periodic orbits controlled by time-delay feedback", *Phys. Lett.*, A 210, 87-94.

Boccaletti, S. and F.T. Arecchi, (1995), "Adaptive control of chaos", *Europhys. Lett.*, 31, 127-132.

Boccaletti, S., D. Maza, H. Mancini, R. Genesio, F.T. Arecchi, (1997), "Control of defects and space-like structures in delayed dynamical systems", *Phys. Rev. Lett.*, 79, 5246-5249.

Braiman, Y. and I. Goldhirsch, (1991), "Taming chaotic dynamics with weak periodic perturbations", *Phys. Rev. Lett.*, 66, 2545-2548.

Grassberger, P. and I. Procaccia, (1983), "Characterization of strange attractors", *Phys. Rev. Lett.*, 50, 346-349.

Hunt, E.R., (1991), "Stabilizing high-period orbits in a chaotic system: the diode resonator", *Phys. Rev. Lett.*, 67, 1953-1955.

Jackson, E.A. and A. Hübler, (1990), "Periodic entrainment of chaotic logistic map dynamics", *Physica D*, 44, 407-420.

Lima, R. and M. Pettini, (1990), "Suppression of chaos by resonant parametric perturbations", *Phys. Rev. A*, 41, 726-783.

Lorenz, E,N., (1963), "Deterministic non-periodic flow", *J. Atmos Scie.*, 20, 130-141.

Mackey, M.C. amd L. Glass, (1977), "Oscillation and chaos in physiological control system", *Science*, 197, 287-292.

Meucci, R., W. Gadomski, M. Ciofini, F.T. Arecchi, (1994), "Experimental control of chaos by means of weak parametric perturbations", *Phys. Rev. E*, 49, R2528-R2531.

Meucci, R., M. Ciofini, R. Abbate, (1996), "Suppressing chaos in lasers by negative feedback", *Phys. Rev. E,* 53, R5537-R5540.
Ott., E., C. Grebogi, A.J. Yorke, (1990), "Controlling chaos", *Phys. Rev. Lett.,* 64, 1196-1199.
Pecora, L.M. and T.L. Carrols, (1990), "Synchronization in chaotic systems", *Phys. Rev. Lett.,* 64, 821-824.
Petrov, V., V. Gaspar, J. Masere, K. Showalter, (1993), "Controlling chaos in Belousov-Zhabotinsky reaction", *Nature,* 361, 240-243.
Plapp, B.B. and A.W. Hübler, (1990), "Nonlinear resonances and suppression of chaos in rf-biased Josephson junction", *Phys. Rev. Lett.,* 65, 2302-2305.
Pyragas, K., (1995), "Control of chaos via extended delay feedback", *Phys. Lett. A,* 206, 323-330.
Pyragas, K., (1992), "Continuous control of chaos by self-controlling feedback", *Phys. Lett. A,* 170, 421-428.
Rossler, O.E., (1976), "An equation for continuous chaos", *Phys. Lett.,* 57A, 397-401.
Rossler, O.E., (1979), "An equation for hyperchaos", Phys. Lett., 71A, 155-157.
Roy, R., T.W. Murphy Jr, T.D. Maier, (1992), "Dynamical control of chaotic laser: experimental stabilization of a globally coupled system", *Phys. Rev. Lett.,* 68, 1259-1262.
Sano, M. and Y. Sawada, (1985), "Measurement of the Lyapunov spectrum from a chaotic time series", *Phys. Rev. Lett.,* 55, 1082-1085.
Singer, J., Y.Z. Wang, H.H. Baum, (1991), "Controlling a chaotic system", *Phys. Rev. Lett.,* 66, 1123-1125.
Theiler, J., B. Galdrikian, A. Longtin, S. Eubank, J.D. Farmer, (1992), "Using surrogate data to detect nonlinearity in time series" in: *"Nonlinear modeling and forecasting",* M. Casdagli and S. Eubank Eds., Addison-Wesley, Reading, MA., pp. 163-188.

CONTROLLING CHAOS WITH APPLICATIONS TO NONLINEAR DIGITAL COMMUNICATION

CELSO GREBOGI

Institüt für Theoretische Physik und Astrophysik
Universität Potsdam, PF 601553, D14415 Potsdam, Germany
and
Institute for Plasma Research and Department of Mathematics
University of Maryland, College Park, Maryland 20742, USA
grebogi@chaos.umd.edu

YING-CHENG LAI

Department of Physics and Astronomy
Department of Mathematics
The University of Kansas
Lawrence, Kansas 66045
lai@poincare.math.ukans.edu, Fax: (785)864-5262

SCOTT HAYES

U. S. Army Research Laboratory
Adelphi, Maryland 20783
hayes@lamp0.arl.mil

ABSTRACT

This paper addresses two related issues: (1) control of chaos and, (2) controlling symbolic dynamics for communication. For control of chaos, we discuss the idea for realizing desirable periodic motion by applying small perturbations to an accessible parameter of the system. The key observation is that a chaotic attractor typically has embedded densely within it an infinite number of unstable periodic orbits. Since we wish to make only small controlling perturbations to the system, we do not envision creating new orbits with very different properties from the already existing orbits. Thus we seek to exploit the already existing unstable periodic orbits and unstable steady states. Our approach is as follows: We first determine some of the unstable low-period periodic orbits and unstable steady states that are embedded in the chaotic attractor. We then examine these orbits and choose one which yields improved system performance. Finally, we apply small controls so as to stabilize this already existing orbit. For the issue of communication, we describe an experiment verifying that the injection of small current pulses can be used to control the symbolic dynamics of a chaotic electrical oscillator to produce a digital communication waveform.

1. Introduction

Control of chaos using unstable periodic orbits embedded in a chaotic attractor was proposed (Ott et al., 1990). The basic idea is as follows. First one chooses an unstable periodic orbit embedded in the attractor, the one which yields the best system performance according to some criteria. Second, one defines a small region around the desired periodic orbit. Due to ergodicity of the chaotic attractor, a trajectory eventually falls into this small region. When this occurs, small judiciously chosen temporal parameter perturbations are applied to force the trajectory to approach the unstable periodic orbit. This method is extremely flexible because it allows for the stabilization of different periodic orbits, depending on one's needs, for the same set of nominal values of the parameter.

Recently, it has been suggested that the idea of controlling chaos can be used to communicate information (Hayes et al., 1993). This seems natural because much of the fundamental theory for describing chaotic dynamics involves concepts from information theory, a field created in the context of practical communication. Concepts from information theory used in chaos include metric entropy, topological entropy, Markov partitions, and symbolic dynamics. Chaotic systems are often said to evolve randomly, but this randomness can be viewed as being the information entropy. It has been shown that the close connection between the theory of chaotic systems and information theory in a way that is more than purely formal (Hayes et al., 1993).

The purpose of this paper is to review the basic idea of controlling chaos and how this idea may be used in communication. We first illustrate the controlling-chaos idea by using the simple one-dimensional logistic map (Sec. 2). Relevant Issues to be discussed include average time to achieve the control and the effect of noise. We then present a more general controlling-chaos algorithm which is applicable to two-dimensional maps or three-dimensional autonomous ordinary differential equations (Sec. 3). In Sec. 4, we present the method of controlling symbolic dynamics for communication. Experimental results are presented in Sec. 5 to show that the controlling chaos idea can be utilized to cause the symbolic dynamics of a chaotic system to track a prescribed symbol sequence, thus allowing us to encode any desired message in the signal from a chaotic oscillator. This thus demonstrates that the natural complexity of chaos provides a vehicle for information transmission in the usual sense. Conclusions are presented in Sec. 6.

2. Controlling Chaos - A One-Dimensional Example

The basic idea of controlling chaos can be understood by considering the following one-dimensional logistic map, one of the best understood chaotic systems,

$$x_{n+1} = f(x_n, \lambda) = \lambda x_n(1 - x_n), \quad (1)$$

where x is restricted to the unit interval $[0,1]$, and λ is a control parameter. It is known that this map develops chaos via the period-doubling bifurcation route (Feigenbaum, 1978). For $0 < \lambda < 1$, the asymptotic state of the map (or the atttractor of the map) is $x = 0$; for $1 < \lambda < 3$, the attractor is a nonzero fixed point $x_F = 1 - 1/\lambda$; for $3 < \lambda < 1 + \sqrt{6}$, this fixed point is unstable and the attractor is a stable period-2 orbit. As λ is increased further, a sequence of period-doubling bifurcations occurs in which successive period-doubled orbits become stable. The period-doubling cascade accumulates at $\lambda = \lambda_\infty \approx 3.57$, after which chaos can arise.

Consider the case $\lambda = 3.8$ shown in Fig. 1a for which the system is apparently chaotic. An important characteristic of a chaotic attractor is that there exists *an infinite number of unstable periodic orbits embedded within it*. For example, a fixed point $x_F \approx 0.7368$ and a period-2 orbit with components $x(1) \approx 0.3737$ and $x(2) \approx 0.8894$, where $x(1) = f((x(2))$ and $x(2) = f((x(1))$, are shown in Fig. 1a.

Now suppose we want to avoid chaos at $\lambda = 3.8$. In particular, we want trajectories resulting from a randomly chosen initial condition x_0 to be as close as possible to the period-2 orbit shown in Fig. 1a, assuming that this period-2 orbit gives the best system performance. Of course, we can choose the desired asymptotic state of the map to be any of the infinite number of unstable periodic orbits. Suppose that the parameter λ can be finely tuned in a small range around the value $\lambda_0 = 3.8$, *i.e.*, λ is allowed to vary in the range $[\lambda_0 - \delta, \lambda_0 + \delta]$, where $\delta << 1$. Due to the nature of the chaotic attractor, a trajectory that begins from an arbitrary value of x_0 will fall, with probability one, into the neighborhood of the desired period-2 orbit at some later time. The trajectory would diverge quickly from the period-2 orbit if we do not intervene. Our task is to program the variation of the control parameter so that the trajectory stays in the neighborhood of the period-2 orbit as long as the control is present. In general, the small parameter perturbations will be time-dependent. We emphasize that it is important to apply only small parameter perturbations. If large parameter perturbations are allowed, then obviously we can eliminate chaos by varying λ from 3.8 to 2.0 for example. Such a large change is not interesting.

The logistic map in the neighborhood of a periodic orbit can be approximated by a linear equation expanded around the periodic orbit. Denote the target period-m orbit to be controlled as $x(i)$, $i = 1, \cdots, m$, where $x(i + 1) = f(x(i))$ and $x(m+1) = x(1)$. Assume that at time n, the trajectory falls into the neighborhood of component i of the period-m orbit. The linearized dynamics

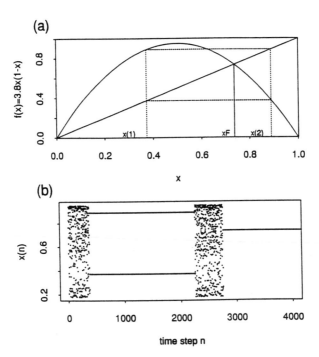

Figure 1: (a) The logistic map $x_{n+1} = f(x_n) = 3.8x_n(1 - x_n)$. An unstable fixed point and an unstable period-2 orbit are also shown. (b) Time series illustrating the control of the period-2 orbit and the fixed point. We choose $\epsilon = 10^{-3}$. The maximum allowed parameter perturbation is $\delta = 5 \times 10^{-3}$.

in the neighborhood of component $i+1$ is then

$$x_{n+1} - x(i+1) = \frac{\partial f}{\partial x}[x_n - x(i)] + \frac{\partial f}{\partial \lambda}\Delta\lambda_n \qquad (2)$$
$$= \lambda_0[1 - 2x(i)][x_n - x(i)] + x(i)[1 - x(i)]\Delta\lambda_n,$$

where the partial derivatives in (2) are evaluated at $x = x(i)$ and $\lambda = \lambda_0$. We require x_{n+1} to stay in the neighborhood of $x(i+1)$. Hence, we set $x_{n+1} - x(i+1) = 0$, which gives

$$\Delta\lambda_n = \lambda_0 \frac{[2x(i)-1][x_n - x(i)]}{x(i)[1-x(i)]}. \qquad (3)$$

Equation (3) holds only when the trajectory x_n enters a small neighborhood of the period-m orbit, i.e., when $|x_n - x(i)| << 1$, and hence the required parameter perturbation $\Delta\lambda_n$ is small. Let the length of a small interval defining the neighborhood around each component of the period-m orbit be 2ϵ. In general, the required maximum parameter perturbation δ is proportional to ϵ. Since ϵ can be chosen to be arbitrarily small, δ also can be made arbitrarily small. As we will see, the average transient time before a trajectory enters the neighborhood of the target periodic orbit depends on ϵ (or δ). When the trajectory is outside the neighborhood of the target periodic orbit, we do not apply any parameter perturbation, and the system evolves at its nominal parameter value λ_0. Hence we set $\Delta\lambda_n = 0$ when $\Delta\lambda_n > \delta$. Note that the parameter perturbation $\Delta\lambda_n$ depends on x_n and is time-dependent.

The above strategy for controlling the orbit is very flexible for stabilizing different periodic orbits at different times. Suppose we first stabilize a chaotic trajectory around the period-2 orbit shown in Fig. 1a. Then we might wish to stabilize the fixed point in Fig. 1a, assuming that the fixed point would correspond to a better system performance at a later time. To achieve this change of control, we simply turn off the parameter control with respect to the period-2 orbit. Without control, the trajectory will diverge from the period-2 orbit exponentially. We let the system evolve at the parameter value λ_0. Due to the nature of chaos, there comes a time when the chaotic trajectory enters a small neighborhood of the fixed point. At this time we turn on a new set of parameter perturbations calculated with respect to the fixed point. The trajectory can then be stabilized around the fixed point.

Figure 1b shows an example where we first control the period-2 orbit and then the fixed point shown in Fig. 1a. The initial condition is $x_0 = 0.28$. At time $n = 381$, the trajectory enters the neighborhood of the component $x(1)$ of the period-2 orbit. For subsequent iterations, the parameter control calculated from (3) is used to stabilize the trajectory around the period-2 orbit. At time

$n = 2200$, we choose to stabilize the trajectory around the fixed point, and hence we turn off the parameter perturbation. The trajectory quickly leaves the period-2 orbit and becomes chaotic. At time $n = 2757$, the trajectory falls into the neighborhood of the fixed point. Parameter perturbations calculated with respect to the fixed point are then turned on to stabilize the trajectory around the fixed point.

In the presense of external noise, a controlled trajectory will occasionally be "kicked" out of the neighborhood of the periodic orbit. If this behavior occurs, we turn off the parameter perturbation and let the system evolve by itself. With probability one the chaotic trajectory will enter the neighborhood of the target periodic orbit and be controlled again. This situation is illustrated in Fig. 2a where we control the period-2 orbit. The noise is modeled by an additive term in the logistic map of the form $\eta\sigma_n$, where σ_n is a Gaussian distributed random variable with zero mean and unit standard deviation, and η is the noise amplitude. The effect of the noise is to turn a controlled periodic trajectory into an intermittent one in which chaotic phases (uncontrolled trajectories) are interspersed with laminar phases (controlled periodic trajectories). It is easy to verify that the averaged length of the laminar phase increases as the noise amplitude decreases, and the length tends to infinity as $\eta \to 0$.

It is interesting to ask how many iterations are required on average for a chaotic trajectory originating from an arbitrarily chosen initial condition to enter the neighborhood ϵ of the target periodic orbit. Clearly, the smaller the value of ϵ, the more iterations that are required. In general, the average transient time $\tau(\epsilon)$ before turning on control scales with ϵ as

$$\tau(\epsilon) \sim \epsilon^{-\gamma}, \qquad (4)$$

where $\gamma > 0$ is a scaling exponent. For one-dimensional maps, the probability that a trajectory enters the neighborhood of a particular component (component i) of the periodic orbit is given by

$$P(\epsilon) = \int_{x(i)-\epsilon}^{x(i)+\epsilon} \rho[(x(i)] dx \approx 2\epsilon\rho[x(i)], \qquad (5)$$

where ρ is the frequency that a chaotic trajectory visits a small neighborhood of the point x on the attractor. We have $\tau(\epsilon) = 1/P(\epsilon) \sim \epsilon^{-1}$, and therefore $\gamma = 1$. This behavior in illustrated in Fig. 2b, where $\tau(\epsilon)$ is plotted on a logarithmic scale for the case of stabilizing the period-2 orbit in Fig. 1a. Twenty values of ϵ were chosen in the range $[10^{-4}, 10^{-2}]$. For each ϵ, we randomly choose 2000 initial conditions (with an uniform probability distribution) and calculate an average transient time. The slope of the straight line is approximately -1.02, indicating good agreement with the theoretical prediction of $\gamma = 1$. For higher

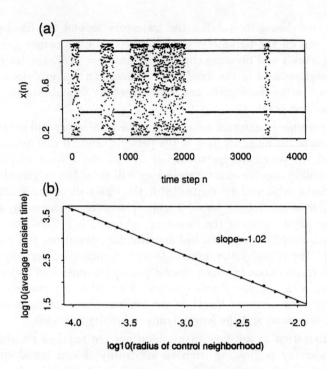

Figure 2: (a) The effect of additive noise modeled by $2.6 \times 10^{-4}\sigma_n$, where σ_n is a Gaussian random variable with zero mean and unit standard deviation. The noise can occasionally kick the controlled trajectory out of the neighborhood of the periodic orbit. (b) Log-log plot of the average time to achieve control $\tau(\epsilon)$ versus ϵ, the size of the controlling neighborhood. The scaling relation between $\tau(\epsilon)$ and ϵ is well fitted by $\tau(\epsilon) \sim \epsilon^{-1}$.

dimensional chaotic systems, the exponent γ can be related to the eigenvalues of the periodic orbit to be controlled (Ott et al., 1990).

A major advantage of the controlling chaos idea (Ott et al., 1990). is that it can be applied to experimental systems in which *a priori* knowledge of the system is usually not known. A time series found by measuring one of the system's dynamical variables in conjunction with the time delay embedding method (Takens, 1981) which transforms a scalar time series into a trajectory in phase space, is sufficient to determine the desired unstable periodic orbits to be controlled and the relevant quantities required to compute parameter perturbations (Ditto et al., 1990; Garfinkel et al., 1992). Another advantage of the method is its flexibility in choosing the desired periodic orbit to be controlled. The method has attracted growing interest for controlling dynamical systems and has been extended to higher dimensional dynamical systems (Romeiras et al., 1992), Hamiltonian systems (Lai et al., 1993b), the control of transient chaos (Tél, 1991) and chaotic scattering (Lai et al., 1993c), and the synchronization of chaotic systems (Lai and Grebogi, 1993). It also has been successfully implemented in various physical experiments (Ditto et al., 1990; Garfinkel et al., 1992).

3. Two-Dimensional Algorithm

The general algorithm for controlling chaos for two dimensional maps (or three dimensional autonomous flows that can be reduced to two dimensional maps on the Poincaré surface of section) can be formulated in a similar way. Consider the two-dimensional map,

$$\mathbf{X}_{n+1} = \mathbf{F}(\mathbf{X}_n, p), \tag{6}$$

where $\mathbf{X}_n \in \mathbf{R}^2$, $\mathbf{F}$ is a smooth function of its variables, and $p \in \mathbf{R}$ is an externally accessible control parameter. We restrict parameter perturbations to be small, i.e.,

$$|p - p_0| < \delta, \tag{7}$$

where p_0 is some nominal parameter value, and $\delta \ll 1$ defines the range of parameter variation. We wish to program the parameter p so that a chaotic trajectory is stabilized when it enters an ϵ-neighborhood of the target periodic orbit. Specifically, let the desired period-m orbit be $\mathbf{X}(1, p_0) \to \mathbf{X}(2, p_0) \to \cdots \to \mathbf{X}(m, p_0) \to \mathbf{X}(m+1, p_0) = \mathbf{X}(1, p_0)$. The linearized dynamics in the neighborhood of component $i+1$ of the period-m orbit is

$$\mathbf{X}_{n+1} - \mathbf{X}(i+1, p_0) = \mathbf{A} \cdot [\mathbf{X}_n - \mathbf{X}(i, p_0)] + \mathbf{B} \Delta p_n, \tag{8}$$

where $\Delta p_n = p_n - p_0$, $\Delta p_n \leq \delta$, $\mathbf{A}$ is a 2×2 Jacobian matrix, and $\mathbf{B}$ is a two dimensional column vector:

$$\begin{aligned} \mathbf{A} &= \mathbf{D}_x \mathbf{F}(\mathbf{X}, p)|_{\mathbf{X}=\mathbf{X}(i), p=p_0}, \\ \mathbf{B} &= \mathbf{D}_p \mathbf{F}(\mathbf{X}, p)|_{\mathbf{X}=\mathbf{X}(i), p=p_0}. \end{aligned} \tag{9}$$

For two-dimensional maps, there exist a stable and an unstable direction at each component of an unstable periodic orbit. The stable (unstable) direction is a direction along which points approach (leave) the periodic orbit exponentially. (For higher dimensional maps, there may be several stable and unstable directions. The algorithm to control chaos in such cases is more complicated (Romeiras et al., 1992) and will not be discussed here.) The existence of both stable and unstable directions at each point of the trajectory can be seen as follows. Choose a small circle of radius ε around an orbit point $\mathbf{X}(i)$. This circle can be written as $dx^2 + dy^2 = \varepsilon^2$ in the Cartesian coordinate system whose origin is at $\mathbf{X}(i)$. The image of the circle under $\mathbf{F}^{-1}$ can be expressed as $A\,dx'^2 + B\,dx'dy' + C\,dy'^2 = 1$, an equation for an ellipse in the Cartesian coordinate system whose origin is at $\mathbf{X}(i-1)$. The coefficients A, B and C are functions of elements of the inverse Jacobian matrix at $\mathbf{X}(i)$. This deformation from a circle to an ellipse means that the distance along the major axis of the ellipse at $\mathbf{X}(i-1)$ contracts as a result of the map. Similarly, the image of a circle at $\mathbf{X}(i-1)$ under $\mathbf{F}$ is typically an ellipse at $\mathbf{X}(i)$, which means that the distance along the inverse image of the major axis of the ellipse at $\mathbf{X}(i)$ expands under $\mathbf{F}$. Thus the major axis of the ellipse at $\mathbf{X}(i-1)$ and the inverse image of the major axis of the ellipse at $\mathbf{X}(i)$ approximate the stable and unstable directions at $\mathbf{X}(i-1)$. We note that typically the stable and unstable directions are not orthogonal to each other, and in rare situations they can be identical (Lai et al., 1993a) (nonhyperbolic dynamical systems).

To calculate these stable and unstable directions, we use an algorithm developed in ref. [12]. This algorithm can be applied to cases where the period of the orbit is arbitrarily large. To find the stable direction at a point $\mathbf{X}$, we first iterate this point forward N times under the map $\mathbf{F}$ and obtain the trajectory $\mathbf{F}^1(\mathbf{X})$, $\mathbf{F}^2(\mathbf{X})$, $\cdots$, $\mathbf{F}^N(\mathbf{X})$. Now imagine we place a circle of arbitrarily small radius ε at the point $\mathbf{F}^N(\mathbf{X})$. If we iterate this circle backward once, the circle will become an ellipse at the point $\mathbf{F}^{N-1}(\mathbf{X})$, with the major axis along the stable direction of the point $\mathbf{F}^{N-1}(\mathbf{X})$. We continue iterating this ellipse backwards, while at the same time rescaling the ellipse's major axis to be order ε. When we iterate the ellipse back to the point $\mathbf{X}$, the ellipse becomes very thin with its major axis along the stable direction at the point $\mathbf{X}$, if N is sufficiently large. For a short period-m orbit, we choose $N = km$ with k an integer. In practice, instead of using a small circle, we take an unit vector at the point $\mathbf{F}^N(\mathbf{X})$, since the Jacobian matrix of the inverse map $\mathbf{F}^{-1}$ rotates a vector in the tangent space of $\mathbf{F}$ towards the stable direction. Hence we iterate a unit vector backward to the point $\mathbf{X}$ by multiplying by the Jacobian matrix of the inverse map at each point on the already existing orbit. We rescale the vector after each multiplication to unit length. For sufficiently large N, the unit vector so obtained at $\mathbf{X}$ is a good approximation to the stable direction at $\mathbf{X}$.

Similarly, to find the unstable direction at point $\mathbf{X}$, we first iterate $\mathbf{X}$ backward under the inverse map N times to obtain a backward orbit $\mathbf{F}^{-j}(\mathbf{X})$ with $j = N, \cdots, 1$. We then choose an unit vector at point $\mathbf{F}^{-N}(\mathbf{X})$ and iterate this unit vector forwards to the point $\mathbf{X}$ along the already existing orbit by multiplying by the Jacobian matrix of the map N times. (Recall that the Jacobian matrix of the forward map rotates a vector towards the unstable direction.) We rescale the vector to unit length at each step. The final vector at point $\mathbf{X}$ is a good approximation to the unstable direction at that point if N is sufficiently large.

The above method is efficient. For instance, the error between the calculated and real stable or unstable directions (Lai et al., 1993a) is on the order of 10^{-10} for chaotic trajectories in the Hénon map if $N = 20$.

Let $e_{s,i}$ and $e_{u,i}$ be the stable and unstable directions at $X(i)$, and let $f_{s,i}$ and $f_{u,i}$ be the corresponding contravariant vectors that satisfy the conditions $f_{u,i} \cdot e_{u,i} = f_{s,i} \cdot e_{s,i} = 1$, and $f_{u,i} \cdot e_{s,i} = f_{s,i} \cdot e_{u,i} = 0$. To stabilize the orbit, we require that the next iteration of a trajectory point, after falling into a small neighborhood about $X(i)$, along the stable direction at $X(i+1, p_0)$, i.e.,

$$[X_{n+1} - X(i+1, p_0)] \cdot f_{u,i+1} = 0. \tag{10}$$

Taking the dot product of both sides of (8) with $f_{u,i+1}$ and use (10), we obtain the following expression for the parameter perturbations,

$$\Delta p_n = \frac{\{A \cdot [X_n - X(i, p_0)]\} \cdot f_{u,i+1}}{- B \cdot f_{u,i+1}}. \tag{11}$$

The general algorithm for controlling chaos for two dimensional maps can be summarized as follows:

1. Find the desired unstable periodic orbit to be stabilized.
2. Find a set of stable and unstable directions, e_s and e_u, at each component of the periodic orbit. The set of corresponding contravariant vectors f_s and f_u can be found by solving $e_s \cdot f_s = e_u \cdot f_u = 1$ and $e_s \cdot f_u = e_u \cdot f_s = 0$.
3. Randomly choose an initial condition and evolve the system at the parameter value p_0. When the trajectory enters the ϵ neighborhood of the target periodic orbit, calculate parameter perturbations at each time step according to (11).

The OGY algorithm described above is not restricted to the control of unstable periodic orbits. The success of the method relies on the existence of distinct stable and unstable directions at trajectory points. It can be applied to stabilizing chaotic trajectories to synchronize two chaotic systems (Lai and Grebogi, 1993) and, consequently, it is also applicable to pseudo-periodic orbits which are chaotic trajectories coming arbitrarily close to some unstable periodic orbits. It should be noted that the OGY algorithm discussed above applies to two-dimensional invertible maps. In general, dynamical systems that can be described by a set of first-order autonomous differential equations are invertible, and the inverse system is obtained by letting $t \to -t$ in the original set of differential equations. Hence, the discrete map obtained on the Poincaré surface of section also is invertible. Most dynamical systems encountered in practice fall into this category. Non-invertible dynamical systems possess very distinct properties from invertible dynamical systems (Chossat and Golubitsky, 1988). For instance, for two-dimensional non-invertible maps, a point on a chaotic attractor may not have a unique stable (unstable) direction. A method for determining all these stable and unstable directions is not known. If one or several such directions at the target unstable periodic orbit can be calculated, the OGY method can in principle be applied to non-invertible systems by forcing a chaotic trajectory to fall on one of the stable directions of the periodic orbit.

4. Application to Nonlinear Digital Communication

The basic strategy for controlling symbolic dynamics for communication is as follows. Imagine that we have a device that can produce chaos (*e.g.*, an electrical oscillator). First, we examine the free-running (i.e., uncontrolled) oscillator and extract from it a symbolic dynamics that allows one to assign symbol sequences to the orbits on the attractor. Typically, some symbol sequences are never produced by the free-running oscillator. The rules specifying allowed and disallowed sequences are called the grammar. Methods for determining the grammar (or an approximation to it) of specific systems have been considered in several theoretical (Cvitanovic *et al.*, 1988; Grassberger *et al.*, 1989) and experimental (Lathrop and Kostelich, 1989) works. (In the engineering literature, a similar concept exists in the context of constrained communication channels.) The next step is to choose a code whereby any message that can be emitted by the information source can be encoded using symbol sequences that satisfy suitable constraints imposed by the dynamics in the presence of the control. It should be noted that the construction of codes with such constraints is a standard problem in information theory (Shannon and Weaver, 1963; Blahut:1988). The code cannot deviate much from the grammar of the free-running oscillator because we envision using only tiny controls that cannot grossly alter the basic topological structure of the orbits on the attractor. Once the code is selected, the next problem is to specify a control method whereby the orbit can be made to follow the symbol sequence of the information to be transmitted. Finally, the transmitted signal must be detected and decoded.

It was described by Hayes *et al.* (1993) how the so-called double scroll electrical oscillator yields a chaotic signal consisting of a seemingly random sequence of positive and negative peaks. If we associate a positive peak with a one, and a negative peak with a zero, the signal yields a binary sequence. We showed how the use of small control perturbations could cause the signal to follow an orbit whose binary sequence represents the information we wish to communicate. Using this method, the chaotic power stage that generates the waveform for transmission can remain simple and efficient (complex chaotic behavior occurs in simple systems), while all the complex electronics controlling the output remains at the low-power microelectronic level.

5. Controlling Symbolic Dynamics for Communication

We now describe the specific experimental procedure used to control the oscillator. Figure 3a is a schematic diagram of the electrical circuit producing the so-called double scroll chaotic attractor, and with tuning of the capacitor C1, the same circuit can be used to produce a single Rössler band (Matsumoto, 1987). The nonlinearity comes from a nonlinear negative resistance represented by the voltage v_R in Fig. 3a. Different realizations of the negative resistance are possible; our circuit synthesizes this negative resistance using an operational amplifier circuit (Matsumoto, 1987). The differential equations describing the double scroll system are

$$C_1 \frac{dv_{C_1}}{dt} = G(v_{C_2} - v_{C_1}) - g(v_{C_1}), \qquad (12)$$
$$C_2 \frac{dv_{C_2}}{dt} = G(v_{C_1} - v_{C_2}) + i_L,$$
$$L \frac{di_L}{dt} = -v_{C_2}.$$

The negative resistance i-v characteristic g is shown in Fig. 3b. For our circuit, we used approximately the component values used by Matsumoto (1987). We take the Poincaré surface of section defined by $v_{C_2} = 0$. Figure 4 shows an experimental trajectory of the double scroll system operating in the Rössler chaos region and the Poincaré surface used. The intersection of the strange attractor with the surface of section is approximately a single straight line segment. We can thus use the value of v_{C_1} alone as a Poincaré coordinate on the surface. To simplify matters for the purpose of real-time computation in the signal processor, we use the quantized value $x = Q[v_{C_1}]$, where Q is the function relating internal quantizer states to the external voltage. Shown also in Fig. 4 are the regions corresponding to symbolic 0 and 1. These regions are separated at the value of v_{C_1} that projects into the fold of the attractor, which is also the peak in the first return map.

To construct a description of the symbolic dynamics of the system, we first run the oscillator without control. When the free-running system state point passes through the surface of section, we record the value of the state-space coordinate x on the surface (corresponding to the quantized value of v_{C_1}), and then record the symbol sequence that is generated by the system after the state point crosses through the surface. Suppose the system generates the binary

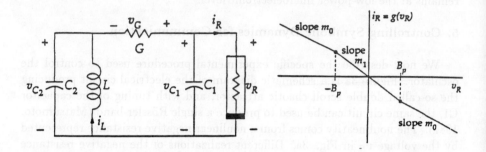

Figure 3: Double scroll oscillator: (a) electrical schematic, and (b) nonlinear negative resistance $i - v$ characteristic g.

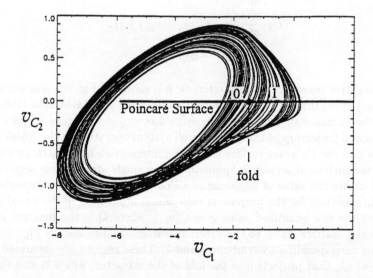

Figure 4: Double scroll oscillator state-space trajectory projected on the $v_{C_1} - v_{C_2}$ plane showing the surface of section.

symbol sequence $b_1 b_2 b_3 \ldots$. We can represent this by the real number $0.b_1 b_2 b_3 \ldots$, so that each symbol sequence corresponds to the real number $r = \sum_{n=1}^{\infty} b_n 2^{-n}$. We refer to the number r, specifying the future symbol sequence, as the symbolic state of the system. This defines a *coding function* mapping the state-space coordinate x on the surface of section to the symbolic coordinate r.

To control the system experimentally using the signal processor, we used a slightly different way to describe the symbolic dynamics. Instead of directly accumulating the coding function, we accumulate an array whose index is a 10-bit symbol sequence, and whose value for a given index is the value of x that *sources* the given symbol sequence. We call this the source statepoint, and denote it by $s[r]$. In this case, r is the *binary value* of the future 10-bit symbol sequence, and not the binary fraction, and ranges from 0 to 1023. Also, so that we can do integer arithmetic in the signal processor for speed, we take s as the raw quantized value of the signal voltage v_{C_1}. This value ranges from about -15,000 to 2000. Now to collect data for $s[r]$, when the statepoint crosses the surface of section, we store the value of x in a temporary variable, and shift the next 10 symbols generated by the natural dynamics of the system into the *symbol* register, thus storing the symbol sequence sourced by x. The value of x is then added to an array $s[r]$, which contains the running sum of statepoints x that source the given symbol sequence r. Dividing by the number of times the system produced the symbol sequence then gives the average value of the source statepoint $s[r]$ for each sequence r. The function $r[s]$ accumulated in this manner is shown in Fig. 5.

To characterize the effects of a control pulse on the symbolic dynamics, we repeat the previous procedure, but now apply a *reference pulse* shortly after the statepoint crosses the surface of section. Now the symbol sequence evolving from a given statepoint will differ from the case when no pulse is applied. Alternatively, we can view the effect of the pulse as being to alter the *statepoint* that *sources* a given symbol sequence, and obtain a description more suited to fast computation of the control pulses. We thus have an array $q[r]$ that is used to store the statepoints that source each possible symbol sequence in the presence of the reference pulses. We applied the control pulse using the digital-to-analog converter (DAC) in our analyzer across a $25 - k\Omega$ resistor connected to the terminal of C_2. We chose this method of coupling because it is simple, and the resistor isolates the signal DAC from the circuit. Power is of course wasted in the resistor, so a more efficient coupling would be used in a device. The application of a $1.5 - V$ reference pulse for $4\mu s$ causes a *differential current* to flow into the circuit, thus altering the dynamics. When this method of coupling is used, current always flows between the DAC and the circuit, so we consider the current difference during the pulse as the differential current provided by the pulse. Note that if a direct coupling is used to inject a current pulse, it is the

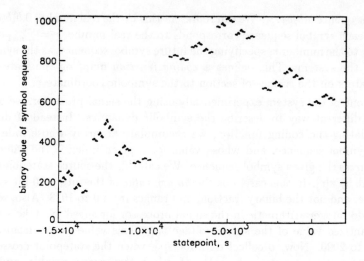

Figure 5: Binary inverse coding function $r[s]$ for the double scroll system when operating in the Rössler region.

differential current that has meaning. The reference pulse causes approximately $60\mu A$ of differential current to flow for about $4\mu s$, thus the average reference current over one cycle (about $200\mu s$) of the oscillator is on the order of a microamp. The precise value of the perturbing current varies depending on when it is applied, the voltage, and other factors that are variable; and this value should not be considered an ultimate limit. A typical root-mean-squared (rms) value of the control pulse is about $0.3V$, thus making the rms differential current on the order of $250nA$. Circuit currents are on the order of $4mA$. The control perturbations are typically much smaller than the reference perturbations and ideally are a zero-mean random process. It is thus difficult to compute the actual control power, because most of the power flow is reactive. We hesitate to view the basic mechanism that is involved here as being amplification of the current pulses by the exponential sensitivity of the dynamics, because ideally the control process is not related to the macroscopic dynamics in any simple way. In an ideal situation, the control pulses would simply cancel the noise in the oscillator that causes its dynamics to drift from a preexisting orbit with the desired symbolic dynamics. Thus, the control process would be a random process with zero mean.

After the arrays s and q have been determined, we produce another array d by the rule $d[r] = p_{ref}/(q[r]-s[r])$ This stored array allows us to quickly compute the required pulse amplitude to correct the symbolic state of the system when the statepoint crosses the surface of section with coordinate x. This correction is given by $p = d[r](x - s[r])$, and thus involves only one integer subtraction and one integer multiply, an operation that can be done very fast in our signal processor. We mention that multipliers exist that can do this at microwave frequencies. Also, for very high speeds, the multiplies could be done with analog circuitry.

We now discuss how we *control* the system to follow a desired binary symbol sequence. Say the system state point passes through the surface of section (shown in Fig. 4) at $x = x_a$, and we want the future symbol sequence to be a given 10-bit sequence r. Because we have previously stored the array $d[r]$ in signal processor memory, we can immediately use the stored value to compute the desired correction pulse. Now, when the system state point crosses the surface of section at $x = x_b$, we have already shifted the new desired bit into the symbol register, and thus have the new desired value of r, and perform the procedure again. On each successive pass through the surface of section, a new code bit is shifted into the symbol register, and we repeat the procedure to correct the symbolic state of the system. The coded information sequence, because it is shifted through the code register, does not begin to appear in the output waveform until N iterations of the procedure, where N is the length of the code register. If the symbol sequence is coming from a properly coded discrete ergodic information source, the process of shifting the information sequence through the code register can be viewed as locking the symbolic dynamics of the oscillator to the information source. Thus, there is a short transient phase during which the symbolic dynamics of the oscillator is being locked to the information source, and the symbolic dynamics of the oscillator is always N bits behind the information source.

Figure 6 shows an experimental information-bearing sequence produced by repeatedly encoding (using a simple binary block coding of the binary digits of the ASCII representation) the quotation "Yea, verily, I say unto you: A man must have Chaos yet within him to birth a dancing star. I say unto you: You have yet Chaos in you." -Friedrich Nietzche, *Thus Spake Zarathustra*. To satisfy the constraints of the grammar in this example we have mapped each binary digit of the ASCII representation of the quotation to two bits according to the rule $1 \rightarrow 11$, $0 \rightarrow 01$. Note that a code producing no more than one zero in a row is sufficient, although the grammar is more complicated. One could also use the variable-length prefix code $1 \rightarrow 1$, $0 \rightarrow 01$, or some other higher-rate code, but the one that we used is sufficient for demonstrating the basic principles. We have not yet constructed a separate receiver circuit yet, but by monitoring the symbol sequence with the analyzer itself and postprocessing, we have verified that the information is encoded correctly. Note that the symbol sequence could be detected from the analog waveform using a very simple thresholding circuit. The fractal nature of the statepoint sequence is caused by the information code

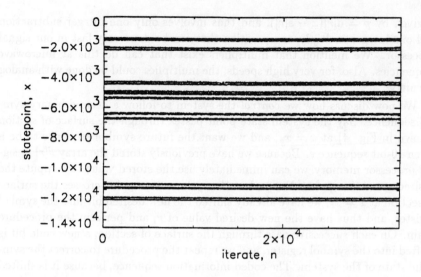

Figure 6: Samples of the controlled signal with the encoded quotation at the Poincaré surface crossing showing the fractal structure of the signal density.

used. Bands in the signal density correspond to finite-length sequences of several binary digits. The fold in the attractor occurs at a quantized value of about -9000, so the symbolic 1 corresponds to the bands ranging from -1000 to -7000, and the symbolic 0 to the region from -11000 to -12000. Subbands correspond to more than one symbol. For example, in the symbolic 0 region, all subbands correspond to the sequence 01, because there are no runs of more than one 0 in a row. The two major sets of subbands in the 01 region, separated by the largest space, thus correspond to the sequences 010 and 011. Thus, if the signal is sampled finely enough, one can extract more than one code bit from one sample.

We now give some general remarks.

1. Since we envision the transmitted signal to be a single scalar, its instantaneous value does not specify the full system state of the chaotic oscillator, and in some cases it might be that such knowledge will be necessary to determine the symbol sequence. If the full system state is needed to extract the symbol sequence, time delay embedding (Tekens, 1981) can be used. As our experiment shows, however, time delay embedding is not always necessary.

2. Because our control technique uses only small perturbations, the dynamical motion of the system is approximately described by the equations for the uncontrolled system. Knowing the equations of motion greatly simplifies the task of removing noise (Kostelich and Yorke, 1988) from a received signal. The

basic nature of the signal in Fig. 6 (symbol regions are separated by a finite distance) implies that the message can still be extracted for noise amplitudes that are significant, but not too large compared to the signal. We have found that signals thus produced have many other interesting properties that are not seen in traditional waveforms.

3. Our control technique can also be used to target a chaotic system in state space. Once the relation between symbolic states and state-space coordinates is established by the coding function $r(x)$, this technique provides a remarkably simple way of doing targeting. We simply shift into the code register the truncated binary string corresponding to the symbolic state r of the desired state-space target x. We have demonstrated that this control method can be used to switch rapidly and without transient chaotic behavior between periodic orbits. Also, we can quickly move from one periodic orbit, say period one, to another, which in effect is subharmonic pulse generation. It may be possible in systems with a stationary state to actually switch the system on and off using small perturbation control, thus forming pulses in the usual sense.

4. Signals that are generated by chaotic dynamical systems and carry information in their symbolic dynamics have an interesting and possibly useful property: More than one encoded symbol can be extracted from a single sample of the trajectory if time delay embedding is used. This can be done in general by using the state-space partition for a higher order iterate of the return map of the system. For the signal produced in our particular experiment, regions mapping to future symbol sequences are physically separated, as shown in Fig. 6.

5. Much of the theory needed to understand information transmission using the symbolic dynamics of chaotic systems already exists (Pecora and Carroll, 1990). For example, because the topological entropy (Adler *et al.*, 1965) of a dynamical system is the asymptotic growth exponent of the number of finite symbol sequences that the system can generate (given the best state-space partition), the channel capacity of a chaotic system used for information transmission is given by the topological entropy.

6. Conclusions

In summary, we have presented an algorithm for converting chaos into periodic motion for one- and two-dimensional maps by using only small parameter perturbations. For two-dimensional maps, the geometrical structure associated with an unstable periodic orbit (namely, the stable and unstable directions) is utilized to effectively achieve the control. In this way the original control method (Ott *et al.*, 1990) can be extended to problems such as controlling chaos in Hamiltonian dynamical systems where unstable periodic orbits always possess complex conjugate eigenvalues (Lai et al., 1993b), synchronizing chaotic systems (Lai and Grebogi, 1993), converting transient chaos into sustained chaos (Lai and Grebogi, 1994), and selecting desirable chaotic states (Nagai and Lai, 1995).

We have also presented an idea of controlling symbolic dynamics of a chaotic system for communication. The idea has been implemented in an experiment using electric circuits. We emphasize that the particular methods for control and coding used in our experiment were chosen for simplicity, and that better methods are possible. Also, the double scroll oscillator itself was chosen because it is simple, and a large body of research is available about its dynamics. It is not intended as an example of a practical oscillator for communication waveform synthesis. It may be possible to use a higher-dimensional radio frequency bandlimited chaotic system for improved performance (higher information rate and better noise immunity), roughly analogous to the use of complex signaling constellations in classical communication systems. We are now developing more practical high-speed symbolic control techniques that could be used at higher bit rates than in our experiment. Finally, we mention that there has been much discussion of the role of chaos in biological systems, and we speculate that the control of chaos with tiny perturbations may be important for information transmission in nature (Shaw, 1981).

Acknowledgements

This work was supported by the Department of Energy (Mathematical, Information and Computational Sciences Division, High Performance Computing and Communication Program). Y.-C. Lai was supported by NSF under Grants No. PHY-9722156 and No. DMS-962659, by AFOSR under Grant No. F49620-96-1-0066, and by the University of Kansas.

References

Adler, R. L., A. G. Konheim and M. H. McAndrew (1965) "Topological entropy", *Trans. Am. Math. Soc.* 114:309.

Blahut, R. E. (1988) *Principles and Practice of Information Theory*, Addison-Wesley.

Chossat, P. and M. Golubitsky (1988) "Symmetry-increasing bifurcation of chaotic attractors", *Physica D* 32:423.

Cvitanovic, P., G. Gunaratne and I. Procaccia (1988) "Topological and metric properties of Henon-type strange attractors", *Phys. Rev. A* 38:1503.

Ditto, W. L., S. N. Rauseo and M. L. Spano (1990) "Experimental control of chaos", *Phys. Rev. Lett.* 65:3211.

Feigenbaum, M. J. (1978) "Quantitative universality for a class of non-linear transformations", *J. Stat. Phys.* 19:25.

Garfinkel, A., M. L. Spano, W. L. Ditto and J. N. Weiss (1992) "Controlling cardiac chaos", *Science* 257:1230.

Grassberger, P., H. Kantz and U. Moenig (1989) "On the symbolic dynamics of the Hénon map", *J. Phys. A* 22:5217.

Hayes, S., C. Grebogi and E. Ott (1993) "Communicating with chaos", *Phys. Rev. Lett.* 70:3031.

Kostelich, E. J. and J. A. Yorke (1988) "Noise reduction in dynamical systems", *Phys. Rev. A* 38:1649.

Lai, Y.-C. and C. Grebogi (1993) "Synchronization of chaotic trajectories using control", *Phys. Rev. E* 47:2357.

Lai, Y.-C, C. Grebogi, J. A. Yorke and I. Kan (1993a) "How often are chaotic saddles nonhyperbolic?", *Nonlinearity* 6:779.

Lai, Y.-C., M. Ding and C. Grebogi (1993b), "Controlling Hamiltonian chaos", *Phys. Rev. E* 47:86.

Lai, Y.-C., T. Tél and C. Grebogi (1993c) "Stabilizing chaotic scattering trajectories using control", *Phys. Rev. E* 48:709.

Lai, Y.-C. and C. Grebogi (1994), "Converting transient chaos into sustained chaos by feedback control", *Phys. Rev. E* 49:1094.

Lathrop, D. P. and E. J. Kostelich (1989) "Characterization of an experimental strange attractor by periodic orbits", *Phys. Rev. A* 40:4028.

Matsumoto, T. (1987) "Chaos in electronic circuits", *Proc. IEEE* 75:1033.

Nagai, Y. and Y.-C. Lai (1995), "Selection of desirable chaotic phase using small feedback control", *Phys. Rev. E* 51:3842.

Ott, E., C. Grebogi and J. A. Yorke (1990) "Controlling chaos", *Phys. Rev. Lett.* 64:1196.

Pecora, L. M. and T. L. Carroll (1990) "Synchronization in chaotic systems", *Phys. Rev. Lett.* 64:821.

Romeiras, F. J., C. Grebogi, E. Ott and W. Dayawansa (1992) "Controlling chaotic dynamical systems", *Physica D* 58:165.

Shannon, C. E. and W. Weaver (1963) *The Mathematical Theory of Communication*, University of Illinois Press.

Shaw, R. (1981) "Strange attractors, chaotic behavior, and information flow", *Z. Naturforsch* 36:80.

Takens, F. (1981) "Detecting strange attractors in turbulence", in: *Lecture Notes in Mathematics*, No. 898, Berlin: Springer-Verlag.

Tél, T (1991), "Controlling transient chaos", *J. Phys. A: Math. Gen.* 24:L1359.

CONTROLLING AND MAINTAINING CHAOS IN BIOLOGICAL SYSTEMS

MARK L. SPANO and VISARATH IN
U.S. Naval Surface Warfare Center, W. Bethesda, Maryland 20817, USA
E-mail: mark@chaos.dt.navy.mil

and

WILLIAM L. DITTO
School of Physics, Georgia Institute of Technology, Atlanta, Georgia 30332, USA

ABSTRACT

Newly devised techniques for the control of chaos in hearts and the control and anti-control of chaos in neural systems are showing promise as viable techniques for the control of dynamical diseases such as ventricular/atrial fibrillation and seizures in brain tissue. Experiments demonstrating the efficacy of control and anti-control of chaos in a rabbit heart septum and a rat hippocampal slice preparations are presented.

1. Introduction

There was a time when scientists believed that simple systems had simple behaviors and complex systems had complex behaviors. They also thought that complex behavior was merely the accumulation of multiple simple behaviors. Twenty years ago these views formed the foundation of our view of nature. Today these illusions have been shattered by the discovery of chaos. We are discovering that nature prizes and exploits change and complexity. The existence of chaos has definitively shown that even simple systems can behave in very complex ways. As usual, nature has exhibited a richer, more elegant structure than our scientific paradigms allowed. From this identification came the recognition that chaos is pervasive in our world (Ditto and Pecora, 1993). It is easy to make simple electronic circuits that are chaotic (Hunt, 1991). Mechanical systems on such wildly different scales as laboratory pendula and orbiting planets have been shown to exhibit chaos (Sussman & Wisdom, 1988). Laser emissions can fluctuate chaotically (Roy *et al.*, 1992). The human heart shows evidence of beating to chaotic rhythms (Garfinkel *et al.*, 1992).

The discovery of chaotic behavior in nature initiated a rapid revolution in the sciences. Chaos has been discovered, accepted and assimilated into the scientific community in the last fifteen years. Until recently, however, it was viewed as a very mathematical and theoretical discipline. More practical people were asking

"What good is chaos?" and answering that it will never be more than a nuisance – something to be avoided in our attempt to use and apply nonchaotic systems. Recently, researchers have demonstrated that chaos admits possibilities and opportunities that simple behavior cannot.

In what follows we will discuss the concept of chaos, from both a theoretical and a laboratory viewpoint, always with an eye towards new methods for the exploitation and control of chaos in biological systems.

Chaos is inevitably confused with randomness and indeterminacy. Because many systems *appeared* random, they were actually thought to *be* random. This was true despite the fact that many of these systems seemed to fall into periods of almost periodic behavior before returning to more "random" motion. Indeed this observation leads to one of the definitions of chaos: the juxtaposition of a very large number of unstable periodic motions. A chaotic system may dwell for a brief time on a motion that is very nearly periodic and then may change to another motion that is periodic with a period that is perhaps five times that of the original motion and so on (Ditto & Pecora, 1993). This constant evolution from one (unstable) periodic motion to another produces a long-term impression of randomness, while showing short-term glimpses of order. These glimpses are not misleading, since chaos is deterministic and not random in nature.

An equivalent definition of chaos is the sensitivity of a chaotic system to small changes in its initial conditions. Thus, if a system is chaotic, these small perturbations quickly (indeed, exponentially) grow until they completely change the behavior of the system. This is both the hope and the despair of those who have to deal with chaotic systems. It is the despair because it effectively renders long term prediction of these systems impossible. Paradoxically, the cause of the despair is also the reason to hope. Because if a system is so sensitive to small changes, could not small changes be used to control it? This realization led Ott, Grebogi and Yorke (OGY) (Ott *et al.*, 1990) to propose an ingenious and versatile method for the control of chaos.

2. The Idea of Control

The starting point for the control of chaos is the phase space of the system. A useful representation of the phase space in biological systems that exhibit discrete events such as beating (in the heart) or electrical bursting (in the brain) can be obtained by plotting the current interval between such events against the previous interval. Such a *Poincaré section* or return plot reduces our information to a manageable level.

By way of contrast, a truly random system will behave in such a way that the points in phase space wander over the entire volume of space available to the system. Its corresponding section will be filled uniformly and densely.

Chaos falls between these two extremes. Since chaos is a superposition of a number of periodic motions, one might expect to see a finite number of points indicating several periodic motions characteristic of the chaotic section. This is true, as far as it goes. However, since chaos is the superposition of a large (*i.e.*, *infinite*) number of periodic motions, the number of points in the section is also infinite. In general, these points form an extended geometric structure, called the system's *chaotic attractor*, which is not a finite set of points (*i.e.*, does not represent periodic motion) and which also does not fill space (*i.e.*, is not random). Generally a chaotic attractor is a fractal object. Knowledge of this attractor and of its response to small perturbations of the system are the only ingredients that are necessary for the control of chaos.

In order to control chaos in biological systems, it is only necessary to identify an unstable periodic point in the attractor, to characterize the shape of the attractor locally around that point and then to put the system onto the stable direction, thereby forcing the system to remain on the unstable motion desired to be stabilized. Let's look at each of these three steps in detail.

The identification of unstable periodic motions is a fairly straightforward process of looking for close returns in the Poincaré section. (Of course it is easier to find points with low order periodicity (small n), but chaos control has been successfully implemented to select periodic motions of order up to about 90!) This is one of the strengths of chaos control: the ability to control on any periodic motion from the infinity of motions present in the system. This capability would allow an engineer to design highly flexible systems employing chaos control.

Characterizing the shape of the attractor is also straightforward. Once we have determined which unstable fixed point we wish to control about, we observe the motion of the point representing the current state of the system (*system state point*) on the attractor. In low dimensional chaotic systems, this point will occasionally approach the vicinity of our chosen unstable fixed point and then move away again. It turns out that, in the neighborhood of the unstable fixed point, the approach is consistently along the same direction (called the *stable direction*) and the departure along another direction (called the *unstable direction*). These two directions, one of which is stable (incoming) and the other of which is unstable (departing), form a saddle around the unstable fixed point (see Figure 1). These *eigenvectors*, along with the speed with which the system state point approaches or departs the vicinity of the unstable fixed point (*stable* or *unstable eigenvalues* respectively), are all that is necessary to characterize the shape of the attractor locally around our chosen point.

The final step for the control of chaos in biological systems is to inject a stimuli to force the system onto the stable direction of the chosen unstable fixed point and thus allow the chaos of the system to exponentially force the motion of the system onto the unstable fixed point.

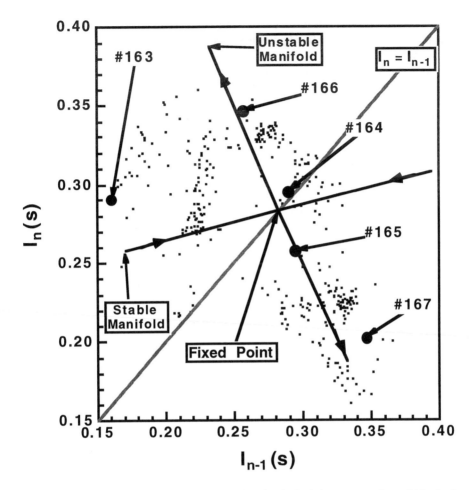

Figure 1 The return map of a typical ouabain-induced arrhythmia heart preparation which clearly demonstrates a local approach and divergence to an unstable fixed point (Garfinkel *et al.*, 1992). Such typical points are candidates for our control of chaos algorithm. I_n is the nth interbeat interval.

3. Heart Experiments

We have implemented chaos control in an experiment in which a section of a rabbit's heart was induced to beat chaotically by the injection of the drug ouabain. The measurable quantity here is the interval between heartbeats, which is about 0.8 sec in the healthy tissue. The effect of the drug is to accelerate the heartbeat and to cause the interbeat intervals to vary chaotically. The chaotic attractor for this with an approach to an unstable fixed point is shown in Figure 1.

To calculate these chaos control interventions it was necessary to use the observed chaotic attractor to predict the time of the next heartbeat. Then an electrical stimulus was used to induce a heartbeat before the predicted natural beat could occur, thereby shortening the next interbeat interval. The amount of time to advance the next beat was calculated so as to place the system state point directly onto the stable direction of the attractor. Thus the succeeding beat would tend to naturally move toward the desired unstable fixed point rather than away from it. This is shown schematically in Figure 2. One should contrast this with on demand pacing in which a simple "no-intervals-above-a-fixed-limit" strategy is employed. During chaos control, stimuli were not applied every cycle, but were only applied as often as necessary to nudge the wandering system onto the stable manifold and hence onto the unstable fixed point. The relevant parameter here is the magnitude of the unstable eigenvalue associated with the unstable fixed point which governs the local rate of expansion along the unstable manifold. Since we could only shorten long beats we were forced to inject control stimuli not every beat but every third or fourth beat as dictated by the presence of a short beat mapping to a long beat along the unstable manifold associated with the desired unstable fixed point. This appears naturally from the control scheme shown in Figure 2.

The results of a typical control run are presented in Figure 3. It was not possible in these experiments to achieve a good period-1 (*i.e.*, "normal") heartbeat. But it was possible to control the chaos consistently into a period 3 beat, which, while not optimal, is better for pumping blood than chaotic beating. As for the relevance to human heart arrhythmias, there is evidence that atrial and ventricular fibrillation may be examples of chaos.

In fact work that is currently underway has shown that human atrial fibrillation is indeed an example of a deterministically chaotic system. Attempts to control this abnormal atrial rhythm have proven successful in converting the irregular, fibrillatory motion of the atrium into periodic motion. Future work is aimed at developing devices that incorporate chaos control algorithms for dealing with these serious and widespread arrhythmias.

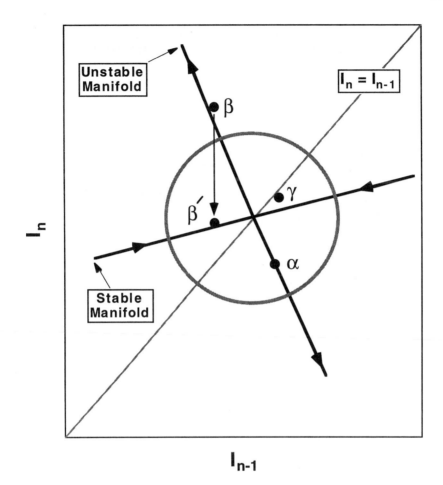

Figure 2 Blowup of the Poincaré map in the region of the unstable fixed point illustrating schematically the idealized chaos control method. Due to ergodicity, the system naturally will sooner or later approach the unstable fixed point. Suppose the system state point is then point α. The system would naturally evolve onto point β on the subsequent beat. However we intervene by shortening the next interbeat interval so that the next point becomes β'. Now the natural dynamics of the system will tend to make the system move toward the unstable fixed point, rather than away from it. Thus the next point is perhaps γ. This will in turn map onto a new α and the cycle begins again. This process will normally produce a period-3 beating, but noise in the system can cause period-2 and period-4 cycles to appear also.

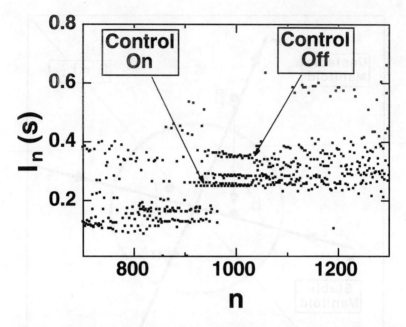

Figure 3 Interbeat interval I_n versus beat number demonstrating the response of the ouabain induced arrhythmia to chaos control interventions (Garfinkel et al., 1992).

4. Brain Experiments

Our success in controlling chaos in the rabbit heart tissue preparation led us to see if a similar strategy could control chaotic behavior in brain tissue. One of the hallmarks of the human epileptic brain during periods in between seizures is the presence of brief burst of focal neuronal activity known as interictal spikes. Often such spikes emanate from the same region of the brain from which the seizures are generated (Gotman, 1991). Several types of *in vitro* brain slice preparations, usually after exposure to convulsant drugs that reduce neuronal inhibition, exhibit population burst-firing activity similar to the interictal spike. One of these preparations is the high potassium [K^+] model, where slices from the hippocampus of the temporal lobe of the rat brain (a frequent site of epileptogenesis in the human) are exposed to artificial cerebrospinal fluid containing high [K^+] which causes spontaneous bursts of synchronized neuronal activity which originate in a region known as the third part of the *cornu ammonis* or CA3 (Korn et al., 1987) as shown in Figure 4.

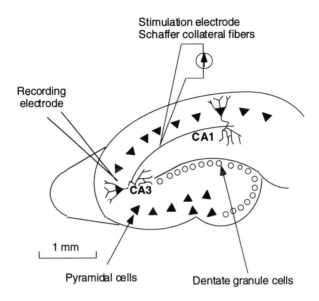

Figure 4 Diagram of transverse hippocampal slice and arrangement of recording electrodes (Schiff et al., 1994).

If one observes the timing of these bursts, clear evidence for unstable fixed points are seen in the return map. Although it was easy to observe the unstable manifolds of these points, it was difficult to accurately determine the stable manifold. This is most likely due to the fact that we are dealing with a high dimensional system whose phase space is being projected onto two dimensions. It may be possible to improve this situation by employing the new high dimensional control techniques described below.

As reported (Schiff et al., 1994) we were able to regularize the timing of such bursts through intervention with stimuli delivered by micropipette with timing as dictated by chaos control to put the system onto the stable direction. As shown in Figure 5, not only were we able to regularize the intervals between spikes but we were also able through an chaos anticontrol strategy to make the intervals more chaotic. It is the latter which might serve a useful purpose in breaking up seizure activity through the prevention or eradication of pathological order in the timing of the spikes.

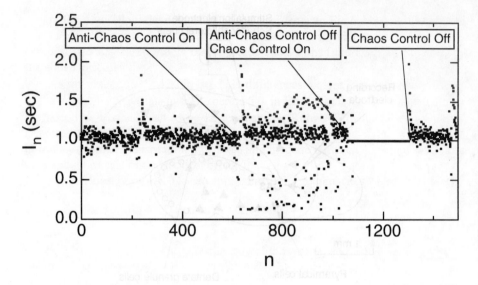

Figure 5 Demonstration of anti-chaos control and chaos control in a hippocampal slice of a rat brain exposed to artificial cerebrospinal fluid containing high [K$^+$] and undergoing spontaneous chaotic population burst firing or spiking.

5. Conclusions

Two fundamental questions dominate future chaos control theories. The first is the problem of controlling higher dimensional chaos. Almost certainly our control and anticontrol of chaotic biological systems will be tremendously enhanced by taking into account the true higher dimensionality of these systems. Just such a higher dimensional chaos control theory has been made by So and Ott (So & Ott, 1995) and a modified version implemented experimentally by Ding and his collaborators (Ding et al., 1996) This method is implementable directly from time series data, regardless of the overall dimension of the phase space.

The second question that has yet to be addressed is the problem of control in a spatiotemporal system. Indeed any true model of biological systems needs to account for the spatial extent of the system. Such systems exhibit both spatial as well as temporal chaos. The study of such systems, while of obvious importance for real-world applications, is as yet in its infancy. The challenge for experimentalists now lies not only in achieving control of higher dimensional and spatiotemporal chaos, but also in guiding the achievements of the past few years out of the laboratory and into our lives.

Acknowledgements

Mark L. Spano and William L. Ditto gratefully acknowledge support from the Office of Naval Research, Physical Sciences Division, as well as from the NSWC Independent Laboratory Internal Research Program (MLS). Visarath In acknowledges support from the ONR/ASEE Postdoctoral Fellowship Program.

References

Ding, M., W. Yang, V. In, W. L. Ditto, M. L. Spano and B. Gluckman, (1996) *Phys. Rev. E* **53**, 4334.

Ditto, W. L. and L. M. Pecora, (1993) *Scientific American*, August, **78**.

Garfinkel, A., M. L. Spano, W. L. Ditto and J. N. Weiss, (1992) *Science* **257**, 1230.

Gotman, J., (1991) *Can. J. Neurol. Sci.* **18**, 573.

Hunt, E. R. (1991) *Phys. Rev. Lett.* **67**, 53.

Korn, S. J., J. L. Giacchino, N. L. Chamberlain and R. J. Dingledine, (1987) *J. Neurophysiol.* **57**, 325.

Ott, E., C. Grebogi and J. A. Yorke, (1990) *Phys. Rev. Lett.* **64**, 1196.

Roy, R., T. W. Murphy, Jr., T. D. Maier and Z. Gills, (1992) *Phys. Rev. Lett.* **68**, 1259.

Schiff, S. J., K. Jerger, D. H. Duong, T. Chang, M. L. Spano and W. L. Ditto, (1994) *Nature* **370**, 615.

So, P. and E. Ott, (1995) *Phys. Rev. E* **51**, 2955.

Sussman, G. J. and J. Wisdom, (1988) *Science* **241**, 433.

SPATIO-TEMPORAL DYNAMICS

SPATIO-TEMPORAL IRREGULARITY OF PROPAGATING ELECTRICAL ACTIVITY IN MODELS OF CARDIAC TISSUE: THE GENESIS OF FLUTTER AND FIBRILLATION

ARUN V. HOLDEN
Department of Physiology and Centre for Nonlinear Studies,
The University, Leeds LS2 9JT, U.K.
E-mail: arun@cbiol.leeds.ac.uk

ABSTRACT

Cardiac excitation equations are stiff, high order and can show bifurcations into chaos. However, nonlinear wave processes - re-entrant waves and their breakdown and destructive interference - underly spatio-temporal irregularity in the arrhythmic heart. The vulnerability to re-entry, and the characteristics of re-entrant waves in 2D atrial and ventricular tissue models are obtained, and their breakdown (predominantly due to rotational anisotropy in the ventricle wall) illustrated.

1. Introduction

The rhythmic beating of the heart is triggered by waves of electrical activity, the propagating cardiac action potentials, that are initiated in a specialised, autorhythmic pacemaker region, and spread through the atria and ventricles through the cardiac tissue, and from the atria to the ventricles by specialised conducting pathways. The rate of beating is controlled by the pacemaker rate, and the pattern of activity arises from propagation through the anisotropic, anatomically organised cardiac muscle. Re-entrant propagation can destroy this rhythmicity, and can break down into lethal fibrillation.

2. ODE models of excitation

Cardiac membrane excitation equations are of the form:

$$\begin{aligned} C\partial_t u &= f(u,v,w) \\ \partial_t v &= g(u,v,w) \\ \partial_t w &= h(u,v,w) \end{aligned} \quad , \quad (1)$$

where $u = u(t)$ is the transmembrane voltage, C is specific membrane capacitance, f is transmembrane current density, vector $v = v(t)$ describes the fast gating variables, and vector $w = w(t)$ comprises slow gating variables and intra-

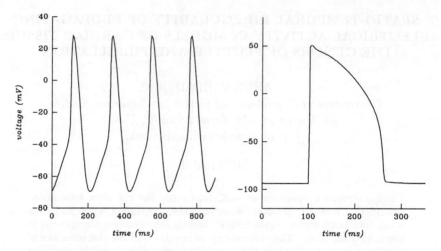

Figure 1: Simulated action potentials for rabbit sinoatrial node: pacemaker activity as a limit cycle trajectory (left) and an evoked action potential in guinea pig ventricular muscle: solution of a stiff ODE. Oxsoft model equations.

and extra-cellular ionic concentrations, and g and h describe their kinetics. The variables u and v have comparable characteristic times.

f can be represented by a simple caricature, such as the FitzHugh-Nagumo (FHN) equations, that retain the cubic characteristics of excitability, or by empirical biophysical excitation equations derived from voltage clamp experiments. The phenomenology of excitation in 2- and 3-dimensional, isotropic and anisotropic excitable media, and in anatomically accurate ventricular wall models, can be explored using FHN models. However, for simulating patterns of propagation in real cardiac tissue, and cardiac arrhythmias and their control, biophysically accurate equations are necessary. Biophysical excitation equations have been obtained for various cardiac tissues; these are high-order, stiff, systems of nonlinear differential equations, and are continually being modified and revised as new experimental observations are incorporated: see chapter 1 of Panfilov & Holden (1996). In this review we use only excitation models from the Oxsoft family, for rabbit sino-atrial node and atrial and guinea-pig ventricular cells (Noble et al., 1994).

The bifurcation behaviour of these high order ODEs can be explored numerically using continuation algorithms, and the common behaviour as a parameter is changed is an equilibrium solution losing, and then regaining, its stability at Hopf bifurcations. Thus the cardiac cell models are either excitable, with a stable equilibrium solution, or endogenously active pacemakers, with a sta-

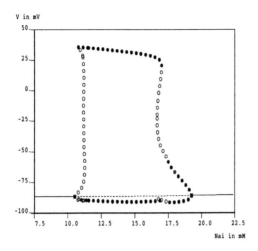

Figure 2: Equilibrium (solid line: stable, dashed line: unstable) and maxima and minima of stable • and unstable o periodic solutions for Oxsoft rabbit atrial cell model as $[Na_i]$ is varied as a bifurcation parameter.

ble limit cycle. Period doubling bifurcations (into bigeminy) have been seen, but patterned and irregular (chaotic) solutions are only found in restricted and unphysiological regions of parameter space.

A simple model of abnormal pacemaker activity induced by ischaemia is to increase $[Na_i^+]$; this simulates the effects of a block of the $Na^+ - K^+$ exchange pump. At elevated $[Na_i^+]$ periodic oscillations emerge: Winslow et al. (1993) have shown that a small cluster of such rhythmically active atrial cells can act as an ectopic pacemaker site, producing an ectopic focus that can trigger an arrhythmia.

3. 1-D PDE models of propagation: vulnerability

A 1-D excitable medium responds to a point excitation by a local subthreshold response, or by a symmetrical pair of travelling waves. If the medium is heterogeneous (say due to its preceeding activity) the wave may only propagate in one direction. In two dimensions this would appear as a wavebreak, and lead to re-entrant propagation, and so the existence of unidirectional propagation is an index of the vulnerability of the tissue to re-entry. The vulnerable window is the time interval after a propagating wave when unidirectional propagation can be initiated. If the stimulated area is large compared to the width of the

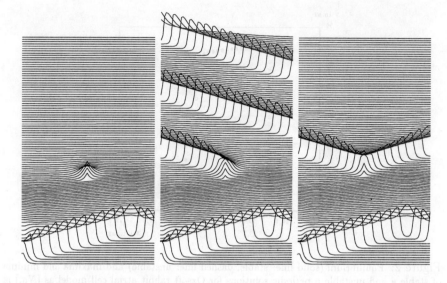

Figure 3: The response of a homogeneous atrial ring model to a test stimulus applied to the refractory tail of a propagating conditioning wave Δt after the conditioning wave front reaches the stimulation site. *left* $\Delta t = 80$ ms, no action potential initiated; *centre* $\Delta t = 83$ ms, stimulus in vulnerable window, unidirectional block; *right*$\Delta t = 90$ ms, a pair of action potential initiated. Membrane potential is displayed against distance.

propagating wave the vulnerable window is of the order of electrode size/wave velocity.

The vulnerable window can be computed for 1-D media with biophysical excitation equations; this provides a computationally efficient method for screening the effects of parameter changes (say changed by pharamacological agents) on the vulnerability to re-entry. Such computations would have demonstrated the arrhythmogenic effects of the "arrhythmia suppressants" - Na^+-channel blockers that suppress ventricular premature depolarisations - used in the CAST trial.

4. 2-D PDE models of propagation: spiral waves

One of the most striking phenomenon in two dimensional excitable media is the existence of reentrant patterns or spiral waves. If we consider a thin ring of excitable tissue, and initiate a single wave which propagates in one direction, then this wave will make a complete turn during the time

$$T = l/V, \qquad (2)$$

where V is the wave velocity, l length of the ring and T the period of rotation. If the period of rotation is longer than the refractory period, this wave will continue to rotate.

Consider a wave front in 2D initially located along a straight line, which starts propagation in one direction around an obstacle. Different points of the front move along different trajectories. The point at the tip of the front rotates around the obstacle, while the point far from the obstacle, moves straight ahead. This results in curving of the initially straight wavefront, and formation of a stationary spiral shape , which will rotate with some period.

For a circular obstacle the spiral shape of the front can be easily determined from simple geometrical arguments as an Archimedean spiral. For cardiac tissue, however, one needs to take into account the anisotropy of cardiac tissue. Such spiral waves can also exist in excitable media without any inexcitable obstacle.

In this case the spiral wave creates a virtual obstacle itself, the core around which it rotates, because of the refractory properties of the medium, and because of curvature effects. The region close to the center of a spiral, the core, determines the main properties of this rotating spiral wave. The simplest example of a core is circular . In this case the tip of a spiral wave rotates along a circle . The rotation of a spiral in this case is stationary, the period of rotation is constant. Such spiral waves usually occur in excitable media with low excitability, when the curvature effects dominates refractoriness. If the excitability of the medium increases, the rotation becomes non-stationary, and meandering of the spiral occurs.

For a single isopotential cell the model is in the form of a system of ordinary differential equations:

$$\begin{aligned} C dV/dt &= f(V, \mathbf{u}) \\ d\mathbf{u}/dt &= \mathbf{g}(V, \mathbf{u}), \end{aligned} \qquad (3)$$

where C is the capacitance of a single cell, V the membrane potential in mV and $\mathbf{u}$ is a vector of the activation and inactivation gating variables and the ionic concentrations that determine the total membrane current $f(V, \mathbf{u})$. This model was incorporated into a partial differential equation model for an excitable medium in the plane (x, y)

$$\begin{aligned} \partial V/\partial t &= C^{-1} f(V, \mathbf{u}) + D\nabla^2 V + F(t) \\ \partial \mathbf{u}/\partial t &= \mathbf{g}(V, \mathbf{u}), \end{aligned} \qquad (4)$$

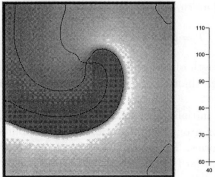 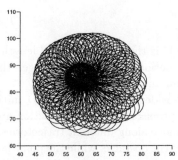

Figure 4: *left* Spiral wave solution displayed as voltage contours every $10mV$ in a 40 mm square rabbit atrial tissue model, the spiral rotates around a circular core, *right* trajectory of tip of spiral wave for first $10s$ of rotation of spiral, aging (due to activity-dependent slow changes in intracellular and extracellular ionic concentrations) changes the pattern of motion from rigid rotation around a small circular core to a biperiodic motion.

where D is the diffusion coefficient for V, ∇^2 is the Laplacian operator $(\partial^2/\partial x^2 + \partial^2/\partial y^2)$

Spiral waves were initiated in two ways, by a cut wavefront or twin pulse protocol. A plane wave was initiated at one edge of the medium by stimulation of a strip and the excitation allowed to propagate to the centre of the medium. The wavefront was then cut, and all the variables on one side of the cut reset to their equilibrium values. This numerically convenient but artificial method allows spirals to be initiated. The twin pulse protocol requires a larger medium, in which a plane wave is initiated at the lower border and later (after the wavefront has propagated through the medium, establishing a gradient in refractoriness) the second stimulus is applied over the left area of the medium.

5. 2-D PDE models of atrial tissue: atrial flutter

Numerical solutions of a 2D atrial tissue model show a rigidly rotating spiral wave solution, that develops a meander after several seconds of rotation - see Holden and Zhang (1994), Biktashev and Holden (1995).

6. 2-D PDE models of ventricular tissue: monomorphic ventricular tachycardia.

The tip of the spiral wave solution may be defined by the intersection of two isolines; for the ventricular tissue model we define the tip by the intersection of

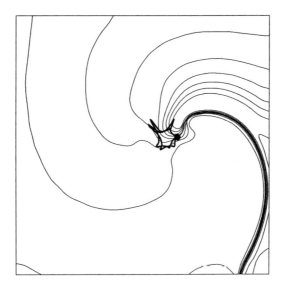

Figure 5: Spiral wave solution of the ventricular tissue model in a 20 × 20 mm medium, as the spatial distribution of isolines of membrane potential V, at an instant 2.67s after the spiral wave was initiated by cutting a broken plane wave. The wavefront of the action potential is where the isolines are close together, and this is where the medium has recovered from previous excitations. The spiral rotated with an initial period of approximately 170ms, and over the first 1s the period decreased to 100–110ms. If the medium were large enough, the distance between successive wavefronts far from the tip, $i.e.$ the wavelength of the spiral would be about 8cm.

the $V = -10$mV and the $f = 0.5$ isolines, where f is now the Ca^{++} inactivation gating variable. The trajectory of the tip of the spiral is not stationary, but meanders, and its motion is nonuniform, moving by a jump-like alternation between fast and very slow phases, with about 5 jumps per full rotation. This motion resembles an irregular, nearly biperiodic process, with the ratio of the two periods close to 1:5. The clear distinction from true biperiodicity, apart from natural experimental errors, can be explained by slow evolution of the parameters of the process, including the two frequencies, and is not a manifestation of dynamic chaos.

The rotation of the spiral wave can be monitored by following an isoline on the wavefront, and the trajectory of the tip of the spiral, as illustrated in Fig. 6. Note that the form of the trajectory here is different from that in Fig. 5; this illustrates ageing of the spiral: wave in Fig. 5 is twice as old as that in Fig. 6, 2.6 s vs 1.3 s. The area enclosed by the tip trajectory is analogous to the core of a rigidly rotating spiral, and is not invaded by the action potential.

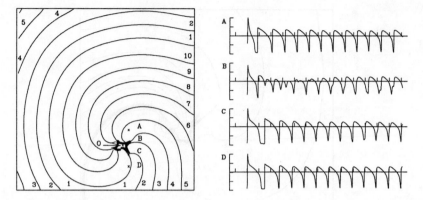

Figure 6: Trajectory of the tip of the spiral wave solution of ventricular tissue model in a 30 × 30 mm medium

Characteristics of the $V(t)$ observed at different sites in the medium during the evolution of a rotating spiral wave are

- only the first action potential, produced by activity invading a resting medium, has the fast depolarisation ($\partial V/\partial t_{\max} > 550\text{V/s}$) and overshoot of 6mV that are characteristic of solitary membrane action potential solutions of (3),

- far from the tip of the spiral wave, the repetitive action potentials are faster and larger than closer to the tip, with $\partial V/\partial t_{\max} > 500\text{V/s}$,

- close to the tip the repetitive action potentials not only are smaller in amplitude, and have reduced $\partial V/\partial t_{\max}$, but appear more as complicated oscillations than action potentials.

Within the core the membrane potential remains between -45 and -5mV; this persistent depolarisation inactivates the inward current i_{Na}, thus blocking propagation into the core.

The multi-lobed pattern of the trajectory of the tip illustrated in Figs. 5 and 6 takes time to develop, and itself develops with time. Figure 7 follows the tip trajectory for spirals initiated from a broken wavefront, and by a twin pulse protocol. In both cases the tip trajectory follows a transient over 3–4 rotations, in which there is an extended "core", with a length of some cms, that evolves into an irregular, biperiodic motion around a core contained within 7mm square. Both the extended transient and the biperiodic motions are composed

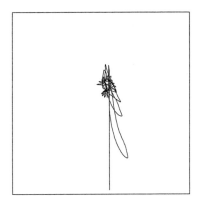 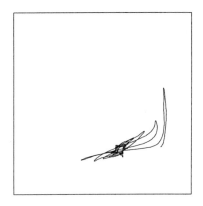

Figure 7: Tip trajectory for ventricular model spiral wave solutions produced by a cut plane wave (left) and a twin stimulus protocol (right), in 30 × 30 *mm* medium

of almost linear segments (when the velocity of the tip movement is fast, about 0.3m/s) broken by sharp turns of up to 170° (when the tip trajectory is almost stationary).

Figure 8 shows that the asymmetric, multilobed trajectory continues to evolve, the area it encloses continues to decrease, and the pattern changes from having two or three sharp turns, to having five. The trajectory during any rotation differs from the trajectory during the preceding rotations.

The spiral wave solutions have been followed for up to 3s, and so, once established, the slowly evolving, almost biperiodic meander pattern shown in Figs 5,6 and 8 appears to be stable.

The spiral wave solutions and the tip trajectories in Figs. 5-8 have all been computed for an isotropic medium, in which the conduction velocity is independent of direction. Ventricular muscle has a laminar structure, with the muscle fibres in a single lamella having an orientation that changes smoothly, and the conduction velocity along the muscle fibres can be up to 3× that transverse to the fibres. This anisotropy in conduction velocity will distort the spiral wave and tip trajectory see Fig.9.

7. Spiral wave breakdown: fibrillation

Spiral wave solutions of some biophysical cardiac excitation equations are unstable (Panfilov & Holden, 1990), and break down into spatio-temporal irregularity. Such spontaneous breakdown of re-entrant activity has been proposed as a mechanism for the development of fibrillation from simple re-entry; the

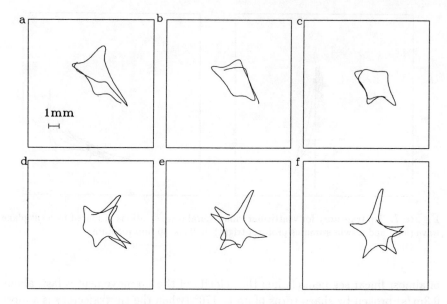

Figure 8: Continuing evolution of tip trajectory of ventricular model

apparent stability of our spiral wave solutions is consistent with the persistence of re-entrant ventricular tachycardia, and suggests that fibrillation may result from geometric *e.g.* rotational anisotropy (Panfilov & Keener, 1993) or intrinsically three-dimensional phenomena *e.g.* negative filament tension (Biktashev et al., 1994).

8. Re-entry in the whole ventricle

A simple method of producing re-entrant waves in two- and three-dimensional models of excitable media is by breaking a planar wavefront, and resetting one side of the wavebreak to a resting state. The broken planar wave develops into a spiral or scroll wave. Figure 10 illustrates the development of re-entry produced by resetting the right half of the ventricle to its resting state: this is a methodological device to generate a broken wavefront, and is not meant to represent a physiological route to re-entry, even though the result (a scroll wave in the ventricles) simulates monomorphic ventricular tachycardia.

The surface view of Fig. 10 shows the scroll wave as a (distorted) spiral: given the geometry of the heart, a re-entrant scroll wave need not appear as a spiral. If the organising filament(s) around which the scroll rotates are deep

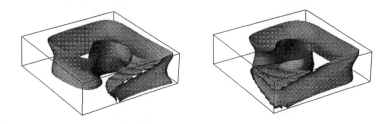

Figure 9: Re-entrant wave in a 5 × 5 × 1.5 mm cuboid 3D model of guinea pig ventricular wall, with 120° transmural rotational anisotropy

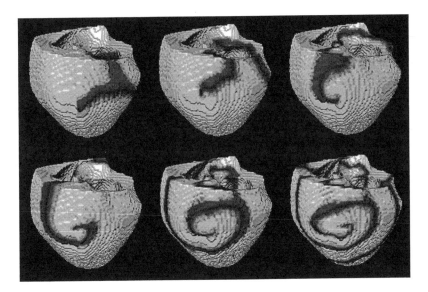

Figure 10: Development of reentrant scroll in canine ventricle CML model - see Holden, Poole & Tucker 1996

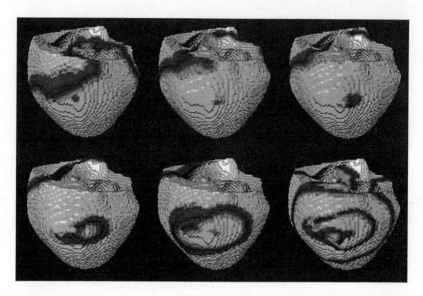

Figure 11: Development of pair of scroll waves in canine ventricle model by stimulation in vulnerable window

in the septum, or the filament is closed as a scroll ring, re-entrant waves can appear on the surface as resembling moving ectopic foci.

An important concept in cardiac electropathology is that the volume of the mammalian myocardial tissue is of the same order of magnitude as the minimum volume that can sustain re-entry. The wavelength of re-entrant action potential (conduction velocity × period) in the canine ventricle is about 5cm, so the parameters used in Fig. 10 scale to a human/pig heart rather than a canine heart.

In numerical simulations it is easy to initiate a spiral by a judicious choice of initial conditions that establishes a broken wavefront. How re-entrant waves arise in cardiac tissue is controversial, and as a stable spiral wave acts an an organising centre it is impossible to determine, from the distribution of voltage in space, how the re-entrant wave was initiated once it is established. Among the mechanisms that can give rise to re-entrant waves in cardiac tissue are the effects of anatomical and functional inhomogeneities and interactions between (almost) planar wavefronts and ectopic sources. Some tissue is more prone to initiate re-entrant arrhythmias–this is characterised by its vulnerability.

Figure 11 illustrates the development of a re-entrant source by a point stimulus in the surface of the model ventricle applied in the vulnerable window during the refractory tail of a wave propagating through the anisotropic tissue.

The rotational anisotropy of the ventricle wall causes the simple scroll waves illustrated above to break down - see Fig. 1 of my accompanying article.

9. Discussion

The relevance of these computations to propagation during re-entrant arrhythmias in ventricular tissue can be assessed by quantitative features of the re-entry — the period and waveform of the action potential, the size of the medium within which re-entry can be initiated, the rapidly decaying transient and then slow ageing of the re-entrant wave (*i.e.* change of period and core shape due to slow processes), the irregular, jump-like motion of the re-entrant wave, and the velocity at which it can be moved by resonant perturbation. The relevance can be questioned, on the grounds that macro- and micro-anatomical detail is ignored, as are the effects of heterogeneity. We would argue that the reasonable correspondence between the computations and observations of re-entrant propagation means that cardiac tissue, in spite of its cellularity, three-dimensional anatomy and heterogeneity, behaves as a reaction diffusion system, and so methods for controlling spiral waves in such systems may be applicable to controlling propagation during re-entrant ventricular arrhythmias.

The transient, extended tip trajectories seen on initiating a spiral wave in Fig. 4 means that although a large medium is required to initiate a reentrant wave, once established, it can survive in a smaller medium. Such a process would allow a re-entrant wave initiated in a transiently vulnerable medium to persist even though the size was not sufficient for re-entry to be initiated in a normal medium *i.e.* a transient increase in vulnerability could allow the creation of a persistent re-entry in a piece of tissue that was smaller then the critical mass for initiating reentry. However, these computations initiate re-entry from a resting medium, not a medium that has been periodically paced at the heart rate, and so even if these transients are seen in two-dimensional epicardial slice preparations they may be an artefact of the method and have no relevance to *in vivo* behaviours

The key characteristic of meander in this ventricular model is the alternation between fast, almost linear, motions and slow, sharp turns. This gives a core with a large perimeter but small area, and results from the dynamic interplay between the fast rate of depolarisation of the action potential forcing high curvature, while the long duration of the action potential only allows small curvatures. This provides a general mechanism for 'linear' conduction blocks in homogeneous excitable media with ventricular-like action potentials. These extended arcs of conduction block would be enhanced by anisotropy in conduction velocity, may give rise to experimental observations interpreted as as a 'linear core'. Unfortunately, existing experimental data do not have enough resolution to distinguish the tiny details of the core, against the background of the visually 'linear' shape caused mainly by the anisotropy.

Simple re-entrant arrhythmias–atrial flutter, or monomorphic ventricular tachycardia–can be identified with a single scroll wave, distorted by anatomy and anisotropy within the atria or ventricles. A possible consequence of such arrhythmias is fibrillation, in which the spatio-temporal activity is irregular. This electrical "cardiac chaos" (in the clinical, not dynamical systems, sense) is manifest as irregular, writhing motion. A possible mechanism for this irregularity is the breakdown of a simple re-entrant wave, due to anisotropy and restitution properties, producing interacting wavefronts that give a maintained spatio-temporal irregularity–an "autowave chaos". Although fibrillation in cardiac tissue is due to more complicated processes, the important point is that such spatio-temporal irregularity can emerge via the effects of anisotropy, in particular, the rotational anisotropy with depth in the myocardial wall, from re-entry in a model or tissue where there are no anatomical or physiological pathologies (i.e. can be a spatio-temporal dynamical disease, resulting from disorders in propagation in structurally and dynamically normal tissue).

Acknowledgement

The work described here was done in collaboration with Dr. Biktashev and supported by grants from the Wellcome Trust(042352, 044365) and the EPSRC ANM initiative (GR/J35641).

References

Biktashev, V.N. & A.V.Holden, (1993). "Resonant drift of an autowave vortex in a bounded medium", *Physics Lett. A* **181**, 216-224.

Biktashev, V.N. & A.V.Holden, (1994). "Design principles for a low voltage cardiac defibrillator based on the effect of feedback resonant drift", *J. Theor. Biol.* **169**, 101-112.

Biktashev, V.N. & A.V.Holden, (1995a). " Resonant drift of autowave vortices in 2*D* and the effects of boundaries and inhomogeneities", *Chaos Solitons and Fractals* **5**, 575-622.

Biktashev, V.N. & A.V.Holden, (1995b). " Control of re-entrant activity in a model of mammalian atrial tissue", *Proc. Roy. Soc. Lond. B* **260**, 211-217.

Biktashev, V.N. & A.V.Holden, (1996). " Re-entrant activity and its control by resonant drift in a two-dimensional model of isotropic homogeneous ventricular tissue" *Proc. Roy. Soc. B* **263** 1373-1383.

Biktashev V.N., A. V.Holden, & H. Zhang (1994). "Tension of organising filaments of scroll waves" *Phil. Trans. Roy. Soc. Lond.* **A347** 611-630.

Holden, A. V., M.J.Poole, & J.V.Tucker (1996). " An algorithmic model of the mammalian heart: propagation, vulnerability, re-entry and fibrillation", *International J. of Bifurcation and Chaos* **6** 1623-1635.

Holden, A. V., & H. Zhang, (1995). "Characteristics of atrial reentry and meander computed from a model of a rabbit single atrial cell", *J. theoretical biology* **175** 545-551.

Keener, J.P. (1996). " Direct activation and defibrillation of cardiac tissue", *J. theoretical biology* **178** 313-324.

Noble, D. (1994). *Oxsoft HEART version 3.4 manual*, Oxsoft: Oxford.

Panfilov, A.V. & A.V.Holden (1990)."Spatio-temporal irregularity in a two-dimensional model of cardiac tissue" *International J. of Bifurcation and Chaos* **1** 219-227.

Panfilov, A.V. & A.V.Holden, (eds) (1997). *The computational biology of the heart* . J.Wiley, Chichester.

Panfilov A.V. & J.Keener (1993). "Re-entry generation in anisotropic twisted myocardium" *J Cardiovasc Electrophysiol* **4** 412-421.

Winslow, R.L. , A.Varghese, D.Noble, C.Adlakha, & A.Hoythya.(1993). " Generation and propagation of ectopic beats induced by spatially localised Na-K pump inhibition in atrial network models" *Proc. Roy. Soc. Lond.* **B 254**, 55-61.

Zipes, D.P. & J.Jalife, (eds) (1995). *Cardiac Electrophysiology - from cell to bedside.* W.B.Saunders, Philadelphia.

DYNAMICAL CONTROL OF BIOLOGICAL RHYTHMS

DAVID J. CHRISTINI
Department of Biomedical Engineering, Boston University
44 Cummington Street, Boston, MA 02215 USA
E-mail: djc@bu.edu

and

JAMES J. COLLINS
Department of Biomedical Engineering, Boston University
44 Cummington Street, Boston, MA 02215 USA
E-mail: collins@enga.bu.edu

ABSTRACT

Quantitative, analytical system models are often unavailable for physiological systems. Thus, traditional model-based control techniques are usually not applicable to the control of biological dynamics. Recently, however, model-independent control techniques have been developed for nonlinear dynamical systems. These techniques do not require explicit knowledge of the underlying system equations and are thus inherently well-suited for the control of biological rhythms. Here we use a model-independent technique to suppress a period-2 rhythm in a cardiac model. This work suggests that model-independent control techniques are applicable to physiological systems. These results may have important clinical implications given that many pathological conditions are characterized by chaotic or nonchaotic dynamics.

1. Introduction

A dynamical system can be defined as a system which continuously changes over time (Strogatz, 1994). The state of a dynamical system is characterized by a set of scalar observables (Thompson & Stewart, 1991). The evolution of such systems can be transient, periodic, or aperiodic. Some dynamical systems (e.g., a forced pendulum) are described by continuous-time differential equations, while others (e.g., the yearly population of an animal species) are described by discrete-time iterated maps, which are also known as difference equations.

Clearly, most physiological systems are dynamical in nature. One example is the cardiovascular system, which can display periodic behavior (e.g., the beating of a healthy heart[1]) or aperiodic behavior [e.g., the beating of a fibrillating heart (Kaplan

---
[1] The beating heart is actually approximately-periodic, as its rate fluctuates slightly about a mean value (Akselrod *et al.*, 1981; Pagani *et al.*, 1986; Christini *et al.*, 1995).

& Cohen, 1990; Hastings *et al.*, 1996; Zeng & Glass, 1996)]. Some dynamical aspects of the cardiovascular system (e.g., the action potential of a single cardiac cell), can be described using differential equations, while others (e.g., the R-R interval time series[2]), require difference equations.

The knowledge that physiological systems are dynamical is not new. For many years, physicians have made diagnoses and prescribed treatments based on the temporal aspects of illnesses. Essentially, physicians have acted as observers who would select a course of action based on the recognition of a pattern in the system's dynamics. In recent years, however, advancements have been made in the understanding of the mechanisms which underlie the dynamical aspects of physiological systems (Mackey & Glass, 1977; Glass & Mackey, 1979; Mackey & Milton, 1987; Guevara *et al.*, 1981; Bélair *et al.*, 1995; Weiss *et al.*, 1994; Glass, 1996). These advances have served to change the emphasis placed on physiological dynamics from observation to understanding.

Along with the understanding of underlying dynamical mechanisms, comes the possibility of exploiting those mechanisms to alter (i.e., control) system behavior. Well-established model-based feedback (i.e., closed loop) control techniques utilize a system's governing equations (i.e., an analytical system model) to control the dynamics of a system (Nise, 1992). Unfortunately, although many physiological mechanisms are well-understood qualitatively, quantitative relationships between system components are usually incomplete. Thus, because accurate analytical system models cannot be developed for such systems, model-based control techniques are currently not applicable to most physiological systems. However, for nonlinear dynamical systems, there is a new class of feedback control techniques for which a qualitative understanding of the underlying mechanisms is sufficient. These new feedback control techniques are model-independent, i.e., they do not require explicit knowledge of a system's underlying equations.[3] Thus, they are applicable to systems that are essentially "black boxes". Model-independent techniques only require qualitative information about component relationships (specifically, the dependence of the variable to be controlled on a system parameter). These techniques extract necessary quantitative system information from an observed time series and then use this extracted information to exploit the system's inherent dynamics to achieve a desired control result. Thus, because qualitative component relationships are the only *a priori* requirement for model-independent control techniques, these techniques are inherently well-suited for the control of physiological systems.

---

[2]The R-R interval is the time between successive electrocardiogram R-waves. The R-R interval is a common method of measuring the inter-beat interval.

[3]For reviews of model-independent control, see Shinbrot *et al.* (1993); Ditto & Pecora (1993); Hunt & Johnson (1993); Lai (1994); Weiss *et al.* (1994); Ott & Spano (1995); Ditto *et al.* (1995); Lindner & Ditto (1995).

2. Model-independent control

In the seminal work in the area of model-independent feedback control, Ott, Grebogi, and Yorke (OGY) (Ott et al., 1990) developed a control technique for chaotic dynamical systems. *Deterministic chaos* can be defined as activity which appears random but actually results from a deterministic, rather than a stochastic, system. This aperiodic behavior occurs indefinitely, because the system never settles down to fixed points, periodic orbits, or quasiperiodic orbits. The chaotic fluctuations are the result of the movement of the system's state point between infinitely many periodic orbits (of varying periods) within a geometrically-finite state space known as a chaotic *attractor*. An attractor is a set of points (in phase space) to which all neighboring trajectories converge (Strogatz, 1994; Devaney, 1989). Once on the attractor, the system trajectory remains there for all time. Because chaotic attractors are ergodic, the system's state point will eventually wander into a given periodic orbit. However, because chaotic systems are highly unstable, the system's state point remains in that orbit only briefly, after which it resumes its aperiodic motion on the chaotic attractor. Thus, because the system's state point cannot (without assistance) remain within the attractor's periodic orbits, these orbits are termed *unstable periodic orbits* (UPOs).

The goal of the OGY control technique is to stabilize one of the infinitely-many UPOs embedded within a chaotic attractor. Once the system's state point wanders into the neighborhood of a desired UPO, the OGY technique repeatedly applies small, time-dependent perturbations to an accessible system parameter. It is of fundamental importance to the OGY control technique that the variable to be controlled is dependent on a system parameter (e.g., drive voltage), which can be adjusted via control perturbations. The first perturbation is designed to force the state point onto the UPO; subsequent perturbations then hold the state point within that UPO. The OGY control technique is depicted schematically in Fig. 1 for a system that can be described by the mapping:

$$x_{n+1} = f(x_n, p) \qquad (1)$$

where x_n is the current value of one measurable system variable, x_{n+1} is the next value of the same variable, and p is an accessible system parameter. The OGY technique utilizes time-delay coordinate embedding (Takens, 1980; Sauer et al., 1991) to define the system's state point in terms of a single measured variable x:

$$\xi_n = [x_n, x_{n+1}]^T. \qquad (2)$$

Information about the attractor,[4] its UPOs, and their stability properties are determined from a first-return map (as shown in Fig. 1). The OGY technique exploits the fact that ξ_n always approaches the unstable periodic fixed point ξ^* ($\xi^* = [x^*, x^*]^T$)

---
[4] Note that the attractor itself is not shown in the schematic of Fig. 1.

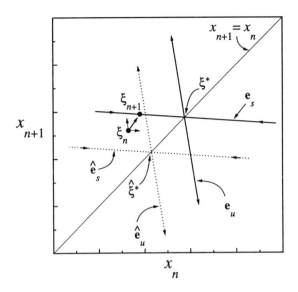

Fig. 1: Geometrical depiction of the OGY control technique. A perturbation made to an accessible parameter shifts the unstable periodic fixed point ξ^* and its stable (e_s) and unstable (e_u) manifolds to $\hat{\xi}^*$, $\hat{e}_s$, and $\hat{e}_u$, respectively. The resultant attracting (due to $\hat{e}_s$) and repelling (due to $\hat{e}_u$) forces move the state point from its current position ξ_n to its next position ξ_{n+1}. At the end of the control perturbation (i.e., when the parameter is returned to its initial value), the unstable periodic fixed point and manifolds return to their original locations, after which the state point ξ_{n+1} is located on e_s. The state point is subsequently attracted towards ξ^* via the stable manifold.

along a characteristic path (the stable manifold) and departs from ξ^* along a different characteristic path (the unstable manifold). Although the system is nonlinear, the stable and unstable manifolds are linear in the neighborhood of ξ^*, where they can be approximated by the unit vectors e_s and e_u, respectively. The control technique perturbs p [to $\hat{p}$ such that the perturbation $\delta p = \hat{p} - p$] such that ξ^*, e_s, and e_u are shifted to $\hat{\xi}^*$, $\hat{e}_s$, and $\hat{e}_u$, respectively. The resultant attracting (due to $\hat{e}_s$) and repelling (due to $\hat{e}_u$) forces move the state point from its current position ξ_n to its next position ξ_{n+1}. At the end of the current control cycle (i.e., when x is re-iterated and p is returned to its initial value), the unstable periodic fixed point and manifolds return to their original locations (ξ^*, e_s, and e_u). The key to the OGY technique is that δp is selected such that the relocated state point ξ_{n+1} will lie on e_s. Following the control perturbation, the state point is subsequently attracted towards ξ^* via the stable manifold. Control is repeated at every iteration to prevent ξ_n from drifting away from ξ^*. Thus, the OGY parameter perturbations constrain the system's state point within the locally-linear UPO neighborhood by exploiting dynamics inherent to that UPO (specifically, the attracting property of the stable manifold). Importantly, ξ^*, e_s, e_u, and g are all estimated from past observations of the system. Thus, the OGY technique is practical from an experimental standpoint because it requires no analytical model of the system.

3. A Technique to Control a Nonchaotic Pathological Rhythm in a Cardiac Model

From a physiological standpoint, model-independent control methods are important because unwanted higher-order oscillations are common in a number of physiological systems. One such pathological higher-order oscillation is known as atrioventricular (AV) nodal conduction time alternans, which is a beat-to-beat alternation (period-2 rhythm) in AV nodal conduction time. The intrinsic rhythm of the mammalian heart is controlled by electrical impulses which originate in a specialized region of cells within the right atrium known as the sinoatrial (SA) node. After leaving the SA node, a cardiac impulse propagates through the atrial tissue to the AV node, which is the only electrical connection between the atria and the ventricles. The impulse then passes through the AV node (with AV nodal conduction time A) and enters the bundle of His, from where it is distributed throughout the ventricular tissue. The AV node is an essential component of cardiac function because it generates a propagation delay which allows ventricular filling and thus facilitates the efficient pumping of blood.

Recently, Sun et al. (1995) developed a model for AV nodal conduction time which accurately reproduced important features (including alternans) of AV nodal rhythms measured experimentally in rabbit hearts. In the model, A_n (the AV nodal conduction time of the nth beat) is a function of H_{n-1} (the AV nodal recovery time following the preceding beat[5]), as given by the following discrete-time difference equation:

$$A_n = A_{min} + S_n + \beta \exp(-H_{n-1}/\tau_{rec}), \qquad (3)$$

where

$$S_1 = \gamma \exp(-H_0/\tau_{fat}), \qquad (4)$$

$$S_n = S_{n-1} \exp[-(A_{n-1} + H_{n-1})/\tau_{fat}] + \gamma \exp(-H_{n-1}/\tau_{fat}), \qquad (5)$$

$$\beta = \begin{cases} 201 \text{ ms} - 0.7 A_{n-1}, & \text{for } A_{n-1} < 130 \text{ ms} \\ 500 \text{ ms} - 3 A_{n-1}, & \text{for } A_{n-1} \geq 130 \text{ ms} \end{cases} \qquad (6)$$

where H_0 is the initial H interval, $A_{min} = 33$ ms, $\tau_{rec} = 70$ ms, $\gamma = 0.3$ ms, and $\tau_{fat} = 30$ s.

---

[5]The AV nodal recovery time is the interval between when one cardiac impulse exits the AV node and the next cardiac impulse enters the AV node.

In Sun et al. (1995), isolated rabbit hearts were electrically stimulated near the SA node at a fixed time interval following bundle of His activation, i.e., the H interval was held constant ($H_n = \overline{H}$). Alternans rhythms were produced when $\overline{H}$ was reduced below a critical value. In the same study, the cardiac model Eqs. (3) – (6) reproduced the *in vitro* alternans rhythms if $\overline{H} < 57$ ms. This stimulation protocol was used as a simple experimental model of re-entrant tachycardia, which can cause AV nodal alternans. A re-entrant tachycardia is a cardiac arrhythmia characterized by a cardiac impulse which passes normally from the atria to the ventricles (via the AV node) but then stimulates another beat when it rebounds back into the atria (via a pathological retrograde pathway) (Simson et al., 1981; Ganz & Friedman, 1995; Glass, 1996). In the above protocol, the shortened H interval was analogous to the shortening caused by a re-entrant impulse.

Figure 2 shows the output of the cardiac model Eqs. (3) – (6) for $\overline{H} = 45$ ms. In Fig. 2, it can be seen that near $n = 250$, the A interval bifurcated into an alternans rhythm. From $1000 \leq n < 1500$, Gaussian white noise ζ_n, with zero mean and standard deviation $\sigma_\zeta = 5$ ms, was added to each H interval ($H_n = \overline{H} + \zeta_n$), causing the A interval to fluctuate. The inset in Fig. 2 shows that there was deterministic structure underlying these fluctuations, i.e., the A intervals ($1430 \leq n < 1449$) in the inset enter the neighborhood of an unstable period-1 fixed point and remain near that unstable period-1 fixed point for several iterates (i.e., beats).

We explored the possibility of exploiting the dynamics depicted in the inset of Fig. 2 to suppress AV nodal alternans in the cardiac model Eqs. (3) – (6) (Christini & Collins, 1996). We implemented a control scheme based on a one-dimensional map simplification (Hunt, 1991; Peng et al., 1991, 1992; Petrov et al., 1994) of the original OGY model-independent chaos control technique. This simplified method applies parameter perturbations[6] which are directly proportional (according to g) to the difference between the system's state point and the nth estimate of the unstable period-1 fixed point $\widehat{A}_n^*$ [see Eqs. (9) and (10) for estimation of $\widehat{A}_n^*$]:

$$\delta H = \frac{A_n - \widehat{A}_n^*}{g}. \tag{7}$$

The proportionality constant g, also known as the fixed-point sensitivity, was estimated as the difference between the unstable periodic fixed point for a given $\overline{H_1}$ and the unstable periodic fixed point for a nearby $\overline{H_2}$ divided by $(\overline{H_1} - \overline{H_2})$:[7]

$$g = \frac{\widehat{A}_{\overline{H_1}}^* - \widehat{A}_{\overline{H_2}}^*}{\overline{H_1} - \overline{H_2}}, \tag{8}$$

---

[6] Here, the parameter perturbation are made to the H interval. Importantly, this parameter is experimentally accessible.

[7] The unstable period-1 fixed point corresponding to each $\overline{H}$ was located by adding Gaussian white noise to each $\overline{H}$, and averaging all A_n that satisfied $|A_n - A_{n-1}| < \epsilon^*$, where ϵ^* defined a small neighborhood around the unstable period-1 fixed point.

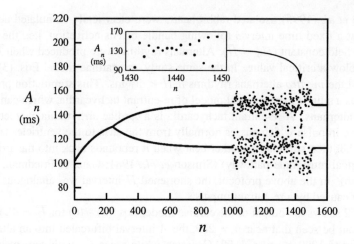

Fig. 2: The AV nodal conduction time A of the cardiac model of Eqs. (3) – (6) for $\overline{H} = 45$ ms. From $1000 \leq n < 1500$, Gaussian white noise ζ_n, with zero mean and standard deviation $\sigma_\zeta = 5$ ms, was added to each H interval ($H_n = \overline{H} + \zeta_n$). A magnified plot of the A intervals for $1430 \leq n < 1449$ is shown in the inset.

where $\widehat{A}^*_{\overline{H}_1}$ and $\widehat{A}^*_{\overline{H}_2}$ are the unstable periodic fixed points for $\overline{H}_1$ and $\overline{H}_2$, respectively.

Figure 3 shows the A and H intervals as functions of beat number n for a representative alternans control trial. The first 1000 points of Fig. 3(a) show an alternans bifurcation produced by $\overline{H} = 50$ ms. At $n = 1000$, model-independent control was activated to eliminate the alternans rhythm. The targeted unstable periodic fixed point was initially estimated as the midpoint of the two A branches:

$$\widehat{A}^*_n = \frac{A_n + A_{n-1}}{2}. \tag{9}$$

Following each H control perturbation, a new unstable periodic fixed point was adaptively estimated as the midpoint of the previous unstable periodic fixed point and the A interval resulting from the control intervention:[8]

$$\widehat{A}^* = \frac{A_n + \widehat{A}^*_{n-1}}{2}, \tag{10}$$

Importantly, the H control perturbations were made to $\overline{H} = 45$ ms. Because this value was less than the "natural" interval of 50 ms, control stimulations were

---

[8] Such adaptive fixed-point estimation is particularly appropriate for biological systems, which are inherently non-stationary.

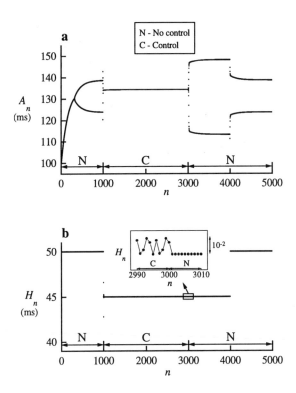

Fig. 3: (a) A and (b) H intervals as functions of beat number n for a representative alternans suppression trial. The respective control stages are annotated in (a) and (b). A magnified plot of the variations in the H interval during the transition between control and no-control is shown in the inset in (b).

not preempted by natural impulses.[9] This control scenario would be appropriate for suppressing an alternans rhythm arising from re-entrant tachycardia, i.e., the control stimulations would not be preempted by re-entrant impulses. Figure 3(a) shows that stabilization of the unstable period-1 fixed point occurred within a few beats and was maintained until adaptive control was turned off at $n = 3000$. A magnified plot of the variations in the H interval during the transition between control and no-control is shown in the inset in Fig. 3(b). From $3000 \leq n < 4000$, H was held constant at its final control value $H_f \approx 45$ ms [Fig. 3(b)]. Due to the unstable nature of the

---

[9]Although lengthening the H interval would be a simple way to eliminate AV nodal alternans, this approach is impractical because inducing beats via electrical stimulation can only shorten the H interval.

period-1 fixed point, the A interval quickly departed from the period-1 rhythm and settled into the stable alternans rhythm corresponding to H_f. At $n = 4000$, $\overline{H}$ was returned to its natural value of 50 ms and the system quickly settled into its original alternans rhythm.

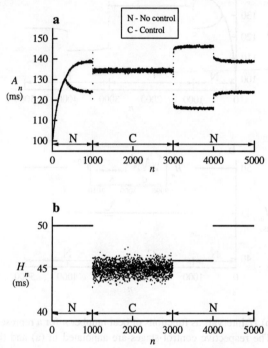

Fig. 4: (a) A and (b) H intervals as functions of beat number n for a representative alternans control trial. Each A iterate, computed using Eq. (3), was "measured" with a precision of 0.2 ms following the addition of measurement noise ($\sigma_\zeta = 0.25$ ms). The respective control stages are annotated in (a) and (b).

Figure 3 shows that model-independent control can be used to suppress alternans in the cardiac model Eqs. (3) – (6). This simulation, however, did not take into account real-world experimental limitations. Two such limitations, imprecise measurements and experimental noise, are inevitable factors associated with biological experiments. Figure 4 shows the results of a control simulation for which each A iterate [computed using Eq. (3)] was "measured" (by the control algorithm) with a precision of 0.2 ms[10] following the addition of a Gaussian "measurement" noise iterate ζ_n ($\sigma_\zeta = 0.25$ ms) to each A_n. Control was robust to the effects of these lim-

---

[10]The same measurement precision was reported in Sun *et al.* (1995).

itations, i.e., the A intervals were successfully constrained within the neighborhood of the unstable period-1 fixed point [Fig. 4(a)].

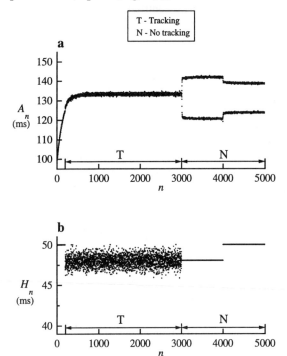

Fig. 5: (a) A and (b) H intervals as functions of beat number n for a representative A-interval tracking trial. Each A iterate, computed using Eq. (3), was "measured" with a precision of 0.2 ms following the addition of measurement noise ($\sigma_\zeta = 0.25$ ms). The respective tracking stages are annotated in (a) and (b).

If model-independent control is initiated prior to the period-doubling bifurcation, the method described above reduces to *tracking* (Carroll et al., 1992; Gills et al., 1992; Schwartz & Triandaf, 1992; Triandaf & Schwartz, 1993). Figure 5 shows a representative A-interval tracking trial with the same limitations that were used for Fig. 4. Tracking was activated at $n = 200$. The period-1 rhythm was effectively stabilized beyond the point ($n \approx 250$) at which the alternans bifurcation was expected. When tracking was turned off at $n = 3000$, the system quickly settled into an alternans rhythm, as in Figs. 3 and 4.

Tracking offers an appealing method for preventing AV nodal alternans if the period-doubling bifurcation into alternans is predictable. However, if it is not known when (or if) the bifurcation will occur, tracking may be impractical given that it may be clinically inappropriate to stimulate a patient's heart before intervention is required. Thus, for an unpredictable alternans bifurcation, a more appropriate approach may be to initiate model-independent control immediately following the bifurcation, thus permitting only a brief period of uncontrolled alternans.

4. Summary

In this study, we have suppressed AV-nodal conduction time alternans in a cardiac model, thus providing evidence that model-independent control techniques can be used to control biological systems. Importantly, the control technique used here is not limited to the suppression of cardiac alternans rhythms. This technique is potentially applicable to any higher-order physiological rhythm in which the variable to be controlled: (1) has an unstable periodic orbit which underlies a stable higher-period orbit, and (2) is directly dependent on an adjustable system parameter. In such a system, our adaptive, model-independent control technique can be used to control the system variable by exploiting the dependency of the UPO on the adjustable parameter. Thus, this technique may be applicable to the control of many physiological dynamical systems. Furthermore, our success in suppressing cardiac alternans may have clinical implications given that alternans in the ECG morphology often precedes life-threatening arrhythmias and is a predictor for sudden death (Smith *et al.*, 1988; Rosenbaum *et al.*, 1994). If a control algorithm similar to the one used in the present study were incorporated into a prosthetic cardiac pacemaker, such alternans rhythms might be suppressed and a route to a fatal arrhythmia curtailed.

References

Akselrod, S., D. Gordon, F. A. Ubel, D. C. Shannon, A. C. Barger & R. J. Cohen (1981) "Power spectrum analysis of heart rate fluctuation: A quantitative probe of beat-to-beat cardiovascular control", *Science* 213:220–222.

Bélair, J., L. Glass, U. an der Heiden & J. Milton (1995) "Dynamical disease: Identification, temporal aspects and treatment strategies of human illness", *CHAOS* 5:1–7.

Carroll, T. L., I. Triandaf, I. Schwartz & L. Pecora (1992) "Tracking unstable orbits in an experiment", *Physical Review A* 46:6189–6192.

Christini, D. J. & J. J. Collins (1996) "Using chaos control and tracking to suppress a pathological nonchaotic rhythm in a cardiac model", *Physical Review E* 53:R49–R52.

Christini, D. J., F. M. Bennett, K. R. Lutchen, H. M. Ahmed, J. M. Hausdorff & N. Oriol (1995) "Application of linear and nonlinear time series modeling to heart rate dynamics analysis", *IEEE Transactions on Biomedical Engineering* 42:411–415.

Devaney, R. L. (1989) *An Introduction to Chaotic Dynamical Systems*, second edition, Addison-Wesley, Reading, Massachusetts.

Ditto, W. L. & L. M. Pecora (1993) "Mastering chaos", *Scientific American* 269:62–68.

Ditto, W. L., M. L. Spano & J. F. Lindner (1995) "Techniques for the control of chaos", *Physica D* 86:198–211.

Ganz, L. I. & P. L. Friedman (1995) "Supraventricular tachycardia", *New England Journal of Medicine* 332:162–173.

Gills, Z., C. Iwata, R. Roy, I. B. Schwartz & I. Triandaf (1992) "Tracking unstable steady states: Extending the stability regime of a multimode laser system", *Physical Review Letters* 69:3169–3172.

Glass, L. (1996) "Dynamics of cardiac arrhythmias", *Physics Today* August:40–45.

Glass, L. & M. C. Mackey (1979) "Pathological conditions resulting from instabilities in physiological control systems", *Annals of the New York Academy of Sciences* 316:214–235.

Guevara, M. R., L. Glass & A. Shrier (1981) "Phase locking, period-doubling bifurcations, and irregular dynamics in periodically stimulated cardiac cells", *Science* 214:1350–1353.

Hastings, H. M., S. J. Evans, W. Quan, M. L. Chong & O. Nwasokwa (1996) "Nonlinear dynamics in ventricular fibrillation", *Proceedings of the National Academy of Science* 93:10495–10499.

Hunt, E. R. (1991) "Stabilizing high-period orbits in a chaotic system: The diode resonator", *Physical Review Letters* 67:1953–1955.

Hunt, E. R. & G. Johnson (1993) "Keeping chaos at bay", *IEEE Spectrum* 30:32–36.

Kaplan, D. T. & R. J. Cohen (1990) "Is fibrillation chaos?", *Circulation Research* 67:886–892.

Lai, Y.-C. (1994) "Controlling chaos", *Computers in Physics* 8:62–67.

Lindner, J. F. & W. L. Ditto, (1995) "Removal, suppression, and control of chaos by nonlinear design", *Appl. Mech. Rev.* 48:795–808.

Mackey, M. C. & L. Glass (1977) "Oscillation and chaos in physiological control systems", *Science* 197:287–289.

Mackey, M. C. & J. G. Milton (1987) "Dynamical diseases", *Annals of the New York Academy of Sciences* 504:16–32.

Nise, N. S. (1992) *Control Systems Engineering*, The Benjamin/Cummings Publishing Company, Inc., Redwood City, California.

Ott, E. & M. Spano (1995) "Controlling chaos", *Physics Today* May:34–40.

Ott, E., C. Grebogi & J. A. Yorke (1990) "Controlling chaos", *Physical Review Letters* 64:1196–1199.

Pagani, M., F. Lombardi, S. Guzzetti, O. Rimoldi, R. Furlan, P. Pizzinelli, G. Sandrone, G. Malfatto, S. Dell'Orto, E. Piccaluga, M. Turiel, G. Baselli, S. Cerutti & A. Malliani (1986) "Power spectral analysis of heart rate and arterial pressure variabilities as a marker of sympatho-vagal ineraction in man and conscious dog", *Circulation Research* 59:178–193.

Peng, B., V. Petrov & K. Showalter (1991) "Controlling chemical chaos", *Journal of Physical Chemistry* 95:4957–4959.

Peng, B., V. Petrov & K. Showalter (1992) "Controlling low-dimensional chaos by proportional feedback", *Physica A* 188:210–216.

Petrov, V., B. Peng & K. Showalter (1994) "A map-based algorithm for controlling low-dimensional chaos", *Journal of Chemical Physics* 96:7506–7513.

Rosenbaum, D. S., L. E. Jackson, J. M. Smith, H. Garan, J. N. Ruskin & R. J. Cohen (1994) "Electrical alternans and vulnerability to ventricular arrhythmias", *New England Journal of Medicine* 330:235–241.

Sauer, T., J. A. Yorke & M. Casdagli (1991) "Embedology", *Journal of Statistical Physics* 65:579–616.

Schwartz, I. B. & I. Triandaf (1992) "Tracking unstable orbits in experiments", *Physical Review A* 46:7439–7444.

Shinbrot, T., C. Grebogi, E. Ott & J. A. Yorke (1993) "Using small perturbations to control chaos", *Nature* 363:411–417.

Simson, M. B., J. F. Spear & E. N. Moore (1981) "Stability of an experimental atrioventricular reentrant tachycardia in dogs", *American Journal of Physiology* 240:H947–H953.

Smith, J. M., E. A. Clancy, C. R. Valeri, J. N. Ruskin & R. J. Cohen (1988) "Electrical alternans and cardiac electrical instability", *Circulation* 77:110–121.

Strogatz, S. H. (1994) *Nonlinear Dynamics and Chaos*, Addison-Wesley, Reading, Massachusetts.

Sun, J., F. Amellal, L. Glass & J. Billette (1995) "Alternans and period-doubling bifurcations in atrioventricular nodal conduction", *Journal of Theoretical Biology* 173:79–91.

Takens, F. (1980) "Detecting strange attractors in turbulence", *In* D. A. Rand & L.-S. Young, eds, *Dynamical Systems and Turbulence (Warwick 1980) (Lecture Notes in Mathematics)*, Vol. 898, Springer, Berlin, pp. 366–381.

Thompson, J. M. T. & H. B. Stewart (1991) *Nonlinear Dynamics and Chaos*, John Wiley and Sons, Chichester, England.

Triandaf, I. & I. B. Schwartz (1993) "Stochastic tracking in nonlinear dynamical systems", *Physical Review E* 48:718–722.

Weiss, J. N., A. Garfinkel, M. L. Spano & W. L. Ditto (1994) "Chaos and chaos control in biology", *Journal of Clinical Investigation* 93:1355–1360.

Zeng, W. & L. Glass (1996) "Statistical properties of heartbeat intervals during atrial fibrillation", *Physical Review E* 54:1779–1784.

THE DYNAMICS OF COUPLED ACTIVE OSCILLATORS

LEONE FRONZONI

Department of Physics, University of Pisa, and INFM, piazza Torricelli 2, Pisa, 56100 Italy
E-mail: fronzoni@difi.unipi.it

ABSTRACT

We consider a population of active oscillators with excitatory or inhibitory all-to-all coupling. This system can mimic the behaviour of natural pacemakers or networks of neurons. By means of analogue simulation it was possible to follow both slow and fast dynamics of the oscillators. At first, we limited our analysis to two or three oscillators in order to understand the basic behaviour due to the coupling and then we extended the analysis to a larger network. Because of the nature of the coupling, the spatial properties are not relevant but we find that the presence of different oscillator frequencies is important to determine the evolution of the population. A wide survey of different dynamics is shown such as phase locking, partially synchronised states, fluctuations and co-operative dynamics of the ensemble.

1. Introduction

In this work we give an experimental analysis of the co-operative dynamics of coupled oscillators. Many papers are present in the literature on this issue and often using a statistical approach (Abbott, 1990; Kuramoto, 1991; Carroll *et al.*, 1994; Sakaguchi, 1994; Ruwisch, 1994; Chia-Chu, 1994; Golumb & Rinzel 1994; Corral *et al.*, 1995; Bottani, 1995; Ernst *et al.*, 1995; Nakagawa & Kuramoto, 1995; Gerstner, 1996). We propose a simple device of low cost that could be made in every laboratory. Of course, many other proposals are present in literature, as for instance, the Chua's Circuit, but I believe that the properties of the device proposed here make it very useful in the study of behaviour of biological elements.

1.1. Non-linear oscillators and biology

Many physiological rhythms are generated either by a single cell or by networks of electrically coupled cells (pacemakers). The self-excited oscillator is the fundamental element that produces rhythms. From the mathematical point of view, this kind of oscillation can be described by a second order differential equation that presents limit cycle or Hopf bifurcation. A classical model is the van der Pol equation that consists of a simple oscillator with non-linear friction term (van der Pol, 1926; van der Pol & van der Mark,1928)

$$\frac{d^2x}{dt^2} = -x + (1-x^2)\frac{dx}{dt}. \quad (1)$$

To understand the limit cycle one has to observe the field of velocities in the space (x,y) described by the following system of equations

$$\frac{dy}{dt} = x, \quad (2)$$

$$\frac{dx}{dt} = -y + x - \frac{x^3}{3}. \quad (3)$$

FIG.1 shows the configuration of this field and the continuous line indicates the limit cycles to which trajectories converge.

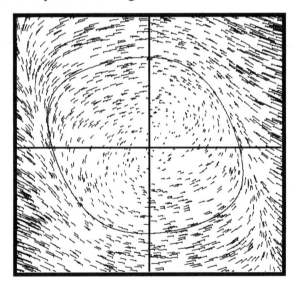

FIG.1. Field of velocity in the Phase portrait of (y, x) variables according to Eq. (1) and Eq. (2). The length of the segments indicates the intensity and its orientation indicates the direction of velocity. The non-uniform configuration of the segments is due to a random choice of the distribution of the points. Continuous line indicates the limit cycle trajectory.

If we add a constant term in Eq. (2) we get

$$\frac{dy}{dt} = x + b \tag{4}$$

This means to impose an asymmetry to the field of velocity. In this case the evolution of the derivative of x(t) assumes a typical pulsed behaviour or firing dynamics as showed in figure.

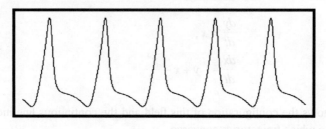

FIG.2. The derivative of x(t) is plotted versus time in the asymmetric conditions (b = 0.8).

1.2 Self-excited oscillator with forcing

When forcing is applied to the non-linear oscillator, synchronisation and chaos are the most interesting properties that it is possible to find. The complexity of the dynamics increases if a forcing is applied to a group of elements.

1.3. Self-excited coupled oscillators

The interest in the phase locking phenomena arising when non-linear oscillators are coupled has recently increased.

Strogatz and Mirollo (1988) suggest a new approach to the study of synchronisation, that consists in limiting the analysis to two coupled oscillators and looking at their phases:

$$\frac{d\phi_1}{dt} = \omega_1 + K \cdot sin(\phi_2 - \phi_1), \tag{5}$$

$$\frac{d\phi_2}{dt} = \omega_2 + K \cdot sin(\phi_1 - \phi_2). \tag{6}$$

By setting

$$\theta = \phi_1 - \phi_2 \quad \text{and} \quad \Delta\omega = \omega_1 - \omega_2 \tag{7}$$

we obtain

$$\frac{d\theta}{dt} = \Delta\omega - 2K\sin\theta, \tag{8}$$

the condition to have phase-locking is

$$\frac{d\theta}{dt} = 0, \tag{9}$$

that is assured when

$$|\Delta\omega| < 2K. \tag{10}$$

This model is unable to describe the dynamics at all, in particular cluster's aggregation, partial synchronisation, transient phase-locking and chaos. Most of the works in the literature are devoted to a statistical description of large network of oscillators. Here we show some special behaviour of a group of self-excited oscillators when global coupling is gradually increased. The oscillators considered are simple electronic devices that are approximately described by the van der Pol equation (1).

The advantage of this technique is widely demonstrated in many works. For instance, the study of L. Pecora and T.L. Caroll (1990) on the synchronisation of chaotic circuits. The possibility of exploring the dynamics by continuously varying the control parameters allowed us to obtain a fast over view of the phenomena. We begin by analysing the effect of the coupling on a system of only two oscillators; then more elements will be considered.

J. Belair and P. Holmes (1984) give a theoretical approach for studying two coupled van der Pol oscillators. They consider linear interactions of the type:

$$\alpha\left(\frac{dy}{dt} - \frac{dx}{dt}\right). \tag{11}$$

To get analytical predictions they approximate a non-linear function by a piecewise linear one. From this analysis they find phenomena of period-doubling bifurcation, chaos and phase-locking. A different approach is to schematize the coupling as a linear function of variables x and y and with asymmetry. In this way the system can better mimic the firing of biological cells. For finite values of b, the oscillator generates spikes very similar to the experimental action potentials.

2. Experimental apparatus

The single oscillator could be realized using a minimum components technique (Fronzoni, 1989), with three operationals and a suitable feedback. The nonlinear friction term is simulate by two diodes connected as shown in FIG.3.

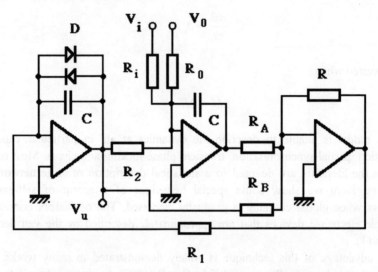

FIG.3. Scheme of the circuit for simulating Eq.(1). V_i means input voltage, V_0 the voltage for controlling asymmetry and V_u is the output corresponding to the x(t) variable. The values of the resistances and capacitances are chosen to reduce the dispersions in the circuits (C = 220 nF or 10 nF, R_0= 680 KΩ, R=R_1=R_2=R_B=R_A= 12 KΩ). The operationals are standard TL082 and diodes (Silicon diode 1N4148).

The output voltage V_u is described by a simple equation of the form

$$C\frac{d^2V_u}{dt^2} = \left[\frac{R}{R_B R_1} - aI_0\left(e^{aV_u} + e^{-aV_u}\right)\right]\frac{dV_u}{dt}$$

$$-\frac{R}{CR_a R_2 R_1}V_u - \frac{R}{CR_a R_1 R_0}V_0 - \frac{R}{CR_a R_1 R_i}V_i \quad (12)$$

The characteristic parameters of the diode provide the values of constants "I_0" and "a". The voltage V_0 fixes the asymmetry of the oscillation. V_0 = 0 means symmetric oscillations. A firing cell is well simulated by means of an asymmetric oscillator or V_0 different from zero. Of course, there is a threshold value below that the oscillator stop to fire, inhibited state. The voltage V_i means the forcing term and in the case of coupling with other oscillators the quantity in front of this means parameter coupling. The resistance R_i assumes the role of coupling resistance and it could be varied in a range of 4 - 200 KΩ.

Developing the exponential terms at the second order this equation takes the known form of the van der Pol equation. In this experiment we have chosen the voltage $V_0 = 1.19$ volt to have pulse-voltage or firing conditions. The global coupling was obtained by connecting the out-put of each element with the input of the others by means of the controlled voltage resistance R_i (H1F1, Isocom Components). In this way it was possible to vary the coupling at the same time for all the oscillators. The parameters are chosen in such a way to have an ordering in frequency but not in the viscosity.

4. Analysis technique

To study the dynamics of the system we used four different techniques:

i) Phase portrait of the global amplitude.
It is particularly useful to get the sum of the amplitudes and its time derivative:

$$Sum = \sum_{j=1}^{N} x_j, \qquad (13)$$

$$Dsum = \sum_{j=1}^{N} \frac{dx_j}{dt}, \qquad (14)$$

where N is the total number of coupled oscillators.

Sending these signals, respectively, to the x-input and y-input of an oscilloscope one gets in real time a projection of the phase-portrait. Closed orbits give a stationary pattern and allow us to distinguish the ordered from the disordered states.

ii) Direct observation of the amplitude oscillations.
By means of a suitable device it is possible to show the amplitude of each oscillator. A stroboscopic image of the amplitudes shows the relative phase between the amplitudes against the presence of no-stationary states.

iii) Fast Fourier Transform.

This traditional method gives information on the presence of periodic trajectories that are indicated by isolated peaks in the spectrum, whereas a continuous profile can indicate the presence of chaos. Special algorithms and non-linear methods are necessary, in this case, as the return maps or the Lyapunov exponents.

iv) Return map and statistical distribution.

This technique allowed us to characterise ordered states against the presence of complex behaviour of the system. The quantity used in this analysis is the time between two successive zero crossing of the global amplitude *Dsum*.

5. Observations and results

5.1. Two coupled oscillators

Two equations describe the dynamics of the systems:

$$\frac{d^2x}{dt^2} = \varepsilon(1-x^2)\frac{dx}{dt} - x + \alpha y, \qquad (15)$$

$$\frac{d^2y}{dt^2} = \varepsilon(1-y^2)\frac{dy}{dt} - x + \alpha x, \qquad (16)$$

The realisation was obtained by connecting the input of an oscillator with the output of the other one. The coupling strength α is an inverse function of the resistance R_i showed in FIG.3 and from Eq. 12 we obtain:

$$\alpha = \frac{R_2}{R_i}\beta. \qquad (17)$$

ε and β are rescaling parameters depending on the diodes characteristics. So we have to distinguish two cases:
a) excitatory coupling (α positive);
b) inhibitory coupling (α negative).

In a recent paper C. van Vreeswijk (1996) makes some general considerations for a population of pulse-coupled oscillators. In the simpe case of two oscillators he shows that the relative phase θ is a function of the sign and value of the coupling parameter α. There is a special value α_c that determines the possible phase between the oscillators.

This result can be resumed in these statements:
excitatory coupling;

$$|\alpha| < |\alpha_c| \Rightarrow \theta = \pi \text{ and } \theta = 0 \text{ are stable,}$$

$$|\alpha| > |\alpha_c| \Rightarrow \theta = \pi \text{ is unstable.}$$

inhibitory coupling;

$$|\alpha| < |\alpha_c| \Rightarrow \theta = 0 \text{ stable and } \theta = \pi \text{ unstable,}$$

$$|\alpha| > |\alpha_c| \Rightarrow \theta = \pi \text{ stable.}$$

5.2. Analysis of firing frequencies and phases

Because the frequency is a relevant parameter for synchronisation it is important to know the behaviour of the frequencies as functions of the coupling parameter. FIG.4 plots the behaviour of frequencies of two oscillators when the strength of excitatory or inhibitory coupling is increased.

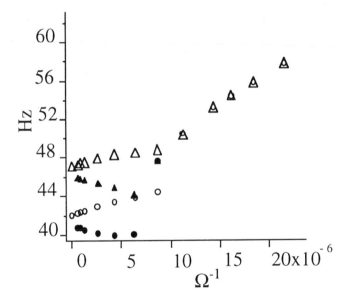

FIG.4. Frequencies of the two coupled oscillators as functions of the reciprocal of the coupling resistance ($1/R_i$). Excitatory coupling (full simbols), inhibitory coupling (open simbols). Circles and triangles correspond to the values of the oscillator frequencies. The phase locking is indicated by the superposition of the values.

In the inhibitory case ($\alpha < 0$), below the threshold ($|\alpha| < |\alpha_c|$), the frequency increases slightly. Over the threshold the data exhibit an increasing, but it is important to stress that the trend in the presence of phase locking depends on the phase difference between the oscillators (FIG.5).

This agrees with the van Vreeswijk's theory that predicts a possible jump to the opposite phase.

As previously observed, the kind of coupling does not determine the trend of the frequency, but we have also to consider the phase relation between the oscillators.

Our inspection of the locked phases showed that:
a) excitatory coupling:
    $\theta = \pi$ $\Rightarrow$ increasing frequency, $\theta = 0$ $\Rightarrow$ decreasing frequency;
b) inhibitory coupling:
    $\theta = \pi$ $\Rightarrow$ decreasing frequency, $\theta = 0$ $\Rightarrow$ increasing frequency.

Next figure still shows the phase locking but with a frequency ratio of 3/2. The triangle's data are the square data multiplied by 3/2. As showed in FIG.6, after the locking threshold triangles superimpose to circles. In this experiment the interaction is inhibitory and $\theta = \pi$.

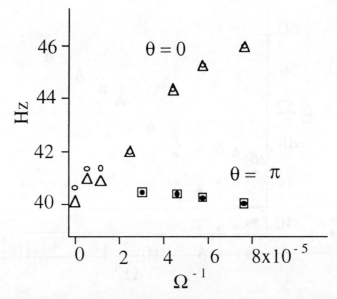

FIG.5. Firing frequency against the reciprocal of coupling resistance R_i (inhibitory case). Phase locking with $\theta = 0$ (increasing trend) and phase locking with $\theta = \pi$ (decreasing trend).

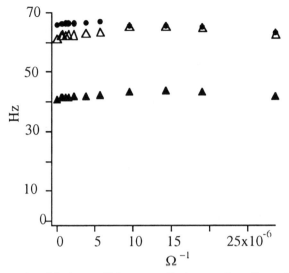

FIG.6. Frequencies of the two oscillators versus the inverse of coupling resistance (full circle and full triangle simbols). Open triangles indicate data obtained multiplying the full triangles data by 3/2. In correspondence of the superimposition of data we observed phase locking between harmonics.

Looking at Eqs. (15) and (16), the observed frequency trend can easily be interpreted. In the presence of phase locking, the amplitudes of both oscillators are the same but with different phases. So, the later equations read

$$\frac{d^2x}{dt^2} = \varepsilon(1-x^2)\frac{dx}{dt} - (1\pm\alpha)x \qquad (18)$$

$$\frac{d^2y}{dt^2} = \varepsilon(1-y^2)\frac{dy}{dt} - (1\pm\alpha)y \qquad (19)$$

The fundamental frequency varies with the coupling strength, as indicated by the term $(1 \pm \alpha)$. The sign of α depends on the kind of coupling and the sign inside the parenthesis depends on differences of phase θ.

5.3. Chaotic behaviour?

It is worth noting that below and near the phase-locking the system becomes noisy but, due to the presence of the electronic noise, it is impossible to assess whether these fluctuations correspond to chaotic dynamics or to a superimposition of complex periodic trajectories. We prefer to speak about "chaotic-like" behaviour.

6. Three coupled oscillators - excitatory mode

With three oscillators the complexities increase further. The dynamics depends on the relative values of frequency. For instance in a condition where two oscillators are locked with opposite phases, the interaction with the third oscillator can produce an inversion of the phases of the first two oscillators. This has been observed for particular values of coupling strength and frequencies. Against the increase of complexity, when more elements are considered, we find some generality:

a) with the inhibitory coupling the locking induces a shift versus high frequency;
b) fluctuations arise below the threshold of locking.

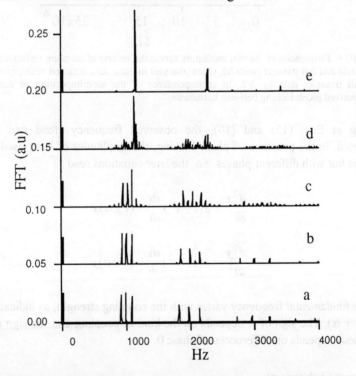

FIG.7. FFT magnitude of the *Sum* data in the case of three oscillators coupled for different values of inhibitory coupling. α/α_g = 0 (a), 0.36 (b), 0.83 (c), 0.983 (d), 1 (e) (α_g is the threshold of global phase-locking).

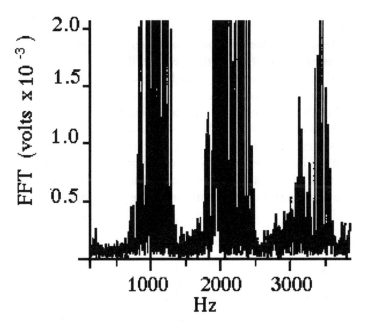

FIG.8. Amplification of the FFT magnitude for case d in FIG.7 (α/α_g = 0.983). The continuous spectrum could imply chaotic dynamics.

This happens also in the case of trapping between harmonic components. By changing the coupling parameter we observe an alternating of ordered and disordered states. FIG.7 shows the FFT magnitude of the global quantity *Sum*. Few isolated peaks are present for low coupling. A shift of the main components to higher frequencies is evident. Near the threshold (row d in FIG.7) the spectrum becomes continuous, as better shown by an amplification of these data in FIG.8.

7. Network of eight oscillators

Increasing the number of oscillators, the number of ordered and noisy states increases. Next figures show the FFT amplitude in the case of eight coupled oscillators. The frequencies of the oscillators were distributed in a range of 500 Hz, around 1200 Hz. See FIG.9 a and FIG.10 a ($\alpha = 0$).

Very complex but periodic behaviour often arise from the chaotic dynamics. The formation of groups or clusters is an interesting phenomenon, particularly evident in FIG.9c where three groups appear near the frequencies of 1000, 1200 and 1400 Hz.

The formation of clusters can be explained both to the small mean average distance of some frequencies and to the drifts of frequency due to the increased coupling, as previously discussed. Figure 9 shows the spectra in the case of inhibitory coupling. The increase of the main frequency to high frequencies and the transition to a continuous spectrum just below the phase locking threshold (case f) are evident. As predicted by Eqs.18,19 the negative sign of α increases the quantity inside the parentheses. As θ is zero on average, this causes a shift towards high frequencies.

Figure 10 shows the FFT magnitude in the case of excitatory coupling. The global effect of the interaction between the oscillators is the formation of clusters with a small shift of the average frequency. This is predicted by Eqs.18,19 considering the negative sign of the quantity inside the parentheses. In this case, the

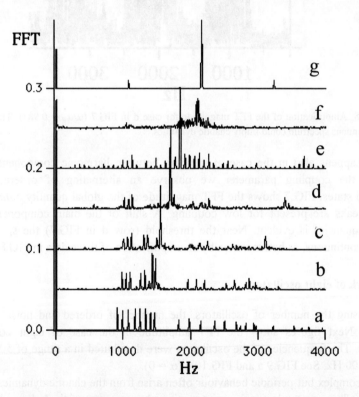

FIG.9. FFT amplitude of the *Sum* series for a system of eight coupled oscillators at different values of inhibitory coupling (α>0). α/α_g = 0 (a), 0.42 (b), 0.54 (c), 0.745 (d), 0.874 (e), 0.999 (f),1 (g). Where α_g is the threshold of global phase-locking.

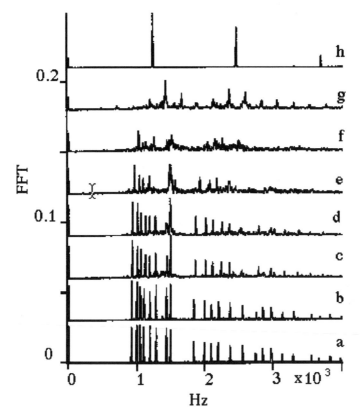

FIG.10. FFT amplitude of the *Sum* series for a system of eight coupled oscillators at different values of excitatory coupling ($\alpha < 0$). $\alpha/\alpha_g =$ 0 (a), 0.27 (b), 37 (c), 0.41 (d), 0.56 (e), 0.73 (f), 0.94 (g), 1 (h).

noisy spectrum appears just below the threshold of phase-locking (case f).

To give a measure of the global dynamics we have computed the integral of the FFT amplitude (IFFTA) as a function of the coupling strength. High values of this quantity mean disordered or complex states. Figure 11 indicates the results for eight coupled oscillators. The coupling strength is indicated in volts according to the values of the voltage applied to the modulated resistance R_i. The jumps to low values of IFFTA correspond to the appearance of periodic orbits that in many cases arise from a locking of different harmonics. A characteristic behaviour is the abrupt increase in complexity, just before the jump to the locked regime as indicated by a

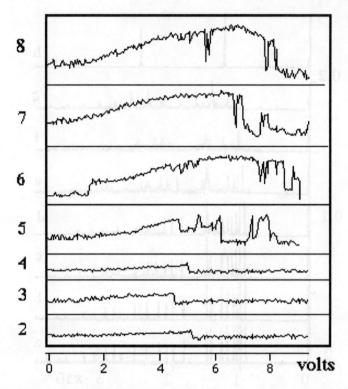

FIG.11. IFFTA as function of the inhibitory coupling. The number on the left indicate the number of coupled oscillators. The coupling strenght is given in voltage applied to the modulating input-resistance R_i.

small peak on the top of the decreasing steps. As the number of oscillators increases the jumps become less evident and the slope of the curve shows a maximum.This behaviour suggests an important consideration about the network dynamics of coupled oscillators. There are two phenomena that, increasing the coupling, are in competition: the increase of the number of bifurcations and the decrease of the dimensions due to phase locking. In other words, at first, the effect of the coupling induces an increase of the periodicity and the complex trajectories in the dynamics of the network, successively, the increase of coupling strength induces phase-locking between the oscillators and so a progressive freezing of the degrees of freedom of the system. There is a special value of coupling at which there is equilibrium between the two processes and this gives a maximum of the number of possible trajectories and a maximum in the IFFTA quantity. As previously discussed, it is particularly hard to decide whether this state is characterised by genuine chaos or a superimposition of periodic trajectories and, also in this case, we prefer to define this as "chaotic-like" behaviour.

An important consideration is that during the "chaotic-like" behaviour the dynamics maintains a sort of memory of the previous ordered state. This is evident by observing, simultaneously, the evolution of the amplitudes of all oscillators on the screen of an oscilloscope. A special device was made for this porpuse that allows us to obtain a stroboscopic image of the oscillators amplitude. It results that against the presence of fluctuations the elements maintain, on average, the relative phase as in the periodic state.

This kind of phenomena is described in the literature and usually refered to as transient phase locking or partial synchronisation. We observed that this dynamics too can generate peaks, in the histogram of the time intervals between two successive zero crossing of global quantity $Dsum$. The results are shown in FIG12 for three values of coupling.

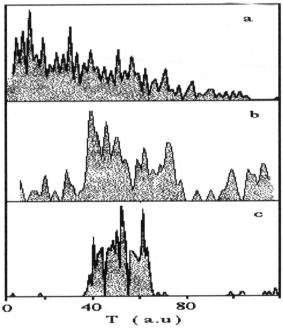

FIG. 12. Histograms of the time intervals between successive zero-crossings of time of $Dsum$ for eight coupled oscillators. (a) Without coupling, $\alpha = 0$. (b) Below the threshold of phase locking, $\alpha = 0.816\ \alpha_g$. (c) Below the threshold and approaching the threshold, $\alpha = 0.944\ \alpha_g$.

This provides us with the explanation of results obtained in some previous experiment on neurons, where the histogram of inter-spike indicates peaked shapes (DeLuca et al., 1993; Carroll, 1995).

The first histogram (FIG.12 a) relates to free oscillators or zero coupling. The next ones (FIG.12 b,c) are performed, respectively, far below and very near the threshold of global locking. Well-separed peaks are present in these cases. In these conditions the phenomenon of transient locking is observable also by means of the stroboscopic devices. From a general point of view this behaviour corresponds to well-defined state of the system where the low-dimensional dynamics is revelated by a structure in the return map. In fact, this is confirmed by the next measurement where the return map of the data previously considered show well-defined structures in presence of transient phase locking or partial synchronisation.

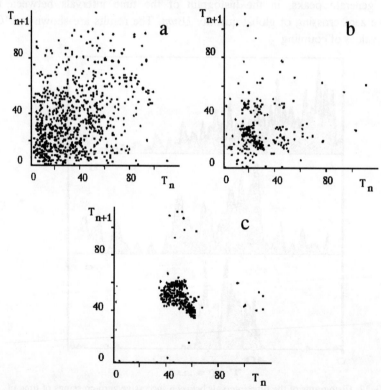

FIG. 13. Return maps of time intervals between successive zero crossing of time of *Dsum* with eight coupled oscillators. Without coupling, $\alpha = 0$ (a), below and approaching the threshold, $\alpha = 0.816\, \alpha_g$ (b), $\alpha = 0.944\, \alpha_g$ (c).

Figure.13a corresponds the case with zero coupling. The points are scattered in a random configuration because of the presence of random phase of the oscillators. The effects of the phase locking induce a reduction of freedom degree of the system and the dynamics collapse to a subspace of the original phase portrait. The low dimensional dynamic is indicated by the collapse of the points in a subset in the return map (FIG.13b,c). The topologic properties of these subsets are at the origin of the multi-peaked histograms of cross-interval of time shown in FIG.12 a,b,c.

8. Conclusions

This paper presents a study of the dynamics of a network of oscillators with limit cycle. We found that the behaviour of the ensemble depends on the relative values of the frequency of the limit cycles. The action of the coupling induces a shift of the frequencies depending on the kind of coupling and on the difference of phase between the elements. The non-linearity due to the coupling induce bifurcation and an increasing of complexity in the dynamics. A numerical evaluation of the global dynamics was obtained by computing the integral of the FFT amplitude (IFFTA) of the global quantity defined by Eqs. 13,14. When the frequency difference of a group of oscillators approaches the threshold of phase locking, the oscillators collapse in a cluster of elements with the same frequency. For small coupling the trend of the IFFTA was increasing and for strong coupling decreasing because of the increasing of the number of synchronization events. It could be particularly interesting to make a prediction of the maximum of the area spectrum as a function of coupling amplitude and, against the evident difficulty that this requires, we will try to elaborate a method to evaluate this threshold. An open problem remains: is the increasing of fluctuations, in proximity of the threshold of phase locking, due to the chaotic dynamics or is it a consequence of internal noise? Our opinion is that this problem is not relevant to the dynamics of the network if compared with the other phenomena as cluster formation, transient synchronization and complexity.

Acknowledgements

I thank M. Barbi, S. Chillemi and A. Di Garbo for their incouragement to make this work and H. Cerdeira for important suggestions and interesting discussions.

References

Abbott, L.F.(1990) "A network of oscillators", *J.Phys.A Math. Gen.* 23:3835-3859.
Belair, J. and P.Holmes (1984) "On linearly coupled relaxation oscillators",
   Quarterly of Applied Mathematics, p. 193.

Bottani, S.(1995)" Pulse-Coupled Oscillators: From Biological Synchronization to Self-Organized Criticality", *Phys. Rev. Lett.* 74:4189.

DeLuca, C.J., A.M. Roy, Z.Erim (1993) "Synchronization of motor-unit firings in several human muscles". *J. Neurophysiol.* 70:20100

Carroll, T.L.(1995) "Synchronization and complex dynamics in pulse-coupled circuit models of neurons", *Biol. Cybern.* 73:553.

Carroll, T.L., J. Heagy, L.M. Pecora (1994) "Synchronization and desynchronization in pulse coupled relaxation oscillators", *Phys. Lett.* A 186:225-229

C. Chia-Chu (1994) " Treshold effects on synchronization of pulse-coupled oscillators", *Phys.Rev.* E 49:2668.

Corral, A., C.J.Pérez, A. Díaz-Guilera, and A. Arenas (1995) " Synchronization in Lattice Model of Pulse-Coupled Oscillators", *Phys. Rev. Lett.* 75:3697

Ernest, U., K. Pawelzik and T. Geisel, (1995) "Synchronization Induced by Temporal Delays in Pulse-Coupled Oscillators", *Phys. Rev. Lett.* 74:1570.

Fronzoni, L. (1989) "Analogue simulations of stochastic processes by means of minimum component electronic devices" in: Noise in Nonlinear Dynamical Systems, F.Moss and P.V.E. McClintock eds., Univ. Press Cambridge, 3:222.

Kuramoto, Y. (1991) "Collective synchronization of pulse-coupled oscillators and exitable units", *Physica D* 50:15-30.

Gerstner, W. (1996) "Rapid Phase-Locking in Systems of Pulse-Coupled Oscillators with Delays", *Phys. Rev. Lett.* 76:1755

Golomb, D. and J. Rinzel (1994) "Clustering in globally coupled inhibitory neurons", *Physica D* 72:259-282.

Nakagawa, N and Y. Kuramoto (1995) " Anomalous Lypunov spectrum in globally coupled oscillators", *Physica D* 80:307-316.

Ruwisch, D., M. Bode, P. Scutz (1994) "Parallel analog computation of coupled cell cycles with electrical oscillators" *Phys. Lett.* A 186:137-144.

Sakaguchi, H. (1994) " Synchronization and a Critical Phenomenon of Pulse-coupled Oscillators", *Prog. Theor. Phys. Lett.* 92:1039.

Strogatz, S.H and R.E. Mirollo "Phase-locking and critical phenomena in lattices of coupled nonlinear oscillators with random intrisic frequencies", *Physica D* 31: 143-168.

van der Pol, B.(1926) "On the relaxation oscillation", *Phil. Mag.* 2:978-92.

van der Pol, B. and van der J. Mark (1928) "The hearthbeat considered as a relaxation oscillation and an electrical model of the heart", *Phil. Mag.* 6:763-75

van Vreeswijk, C.(1996) "Partial synchronization in populations of pulse-coupled oscillators", *Phys. Rev.* E 54:5522.

APPROACHES TO DEFIBRILLATION: THE CONTROL OF RE-ENTRANT ACTIVITY, AND THE ANNIHILATION OF PROPAGATING WAVES OF EXCITATION IN CARDIAC TISSUE

ARUN V. HOLDEN

Department of Physiology and Centre for Nonlinear Studies,
The University, Leeds LS2 9JT, U.K.
E-mail: arun@cbiol.leeds.ac.uk

ABSTRACT

The propagation of electrical activity in cardiac tissue can be modelled by reaction diffusion equations, where a tensor of diffusion coefficients represents anisotropy due to fibre orientation, and excitation is represented by high order, stiff differential systems. The effects of external electrical stimulation, as in artifical pacemakers, or in defibrillators, requires bidomain models, in which intra- and extracellular currents are treated separately. Simplified approaches are taken to this problem to illustrate two methods of defibrillation; by a single large pulse, that eliminates all propagating activity, and by a series of smaller amplitude perturbations, that drive out re-entrant sources of excitation.

1. Introduction

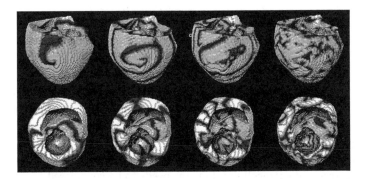

Figure 1: Frames from a movie illustrating the irregular pattern of propagation in the canine ventricles during fibrillation. See Holden, Poole & Tucker, 1966

Disturbances in propagation, or abnormal, ectopic pacemaker sites, can give rise to arrhythmias that can range from being unnoticed to lethal. A dangerous class of arrhythmia are the re-entrant arryhthmias, in which the same wave of excitation repeatedly re-invades the same piece of tissue; these re-entrant arrhythmias are high frequency, as the period of the re-entrant wave is less than the normal period of the heartbeat, and underly atrial flutter and monomorphic ventricular tachycardia. If re-entrant waves break down, due to their intrinsic instability, or the effects of anisotropy and the geometry of the heart, spatio-temporal irregularity in the pattern of activation produces fibrillation, in which different parts of the same chamber of the heart are activated at different times. Global coordination of the contraction of the heart is lost and instead of pumping rhythmically the heart writhes and quivers. The circulation is no longer maintained and death can result if the heart is not defibrillated (Zipes and Jalife, 1995).

Here I consider mathematical aspects of defibrillation by applied electrical shocks or perturbations, where by defibrillation I mean eliminating single and multiple re-entrant wave sources in the heart. The aim is to illustrate the mathematical basis, in terms of reaction-diffusion models for excitable media; important practical technical details, such as electrode geometry and position, and control algorithms, are omitted.

2. PDE models of propagation

For a one dimensional excitable medium with a nonlinear current-voltage relation for the ionic current density I_{ion} (measured in $\mu A cm^{-2}$)

$$C_m \partial V/\partial t = \partial/\partial x (G \partial V/\partial x) + I_{\text{ion}}, \tag{1}$$

where G is a conductance, with units of S. If we divide by C_m we obtain

$$\partial V/\partial t = \partial/\partial x (D \partial V/\partial x) + I_{\text{ion}}/C_m \tag{2}$$

where D is a *diffusion coefficient* with units of $cm^2 ms^{-1}$. Experimental preparations in cardiac electrophysiology are short compared to the spatial extent of a propagating action potential and so a finite cable, with Neumann boundary conditions, is usually appropriate.

To consider the effects of external stimulation the pattern of current flow in both the intra- and extracellular domains needs to be considered - see Chapter 3 of Panfilov and Holden, 1997. Let V_{ex} and V_{int} be potentials and D_{ex} and D_{int} be conductivities in extracellular and intracellular spaces. Assuming Ohmic conductivities, the currents in extracellular and intracellular spaces are given by:

$$I_{ex} = -D_{ex} \frac{\partial V_{ex}}{\partial x} \quad I_{int} = -D_{int} \frac{\partial V_{int}}{\partial x} \tag{3}$$

From the conservation law for current flow :

$$\frac{\partial I_{ex}}{\partial x} = i_m, \quad \frac{\partial I_{int}}{\partial x} = -i_m, \tag{4}$$

The membrane current i_m is a sum of ionic and capacitive current, therefore the equations become:

$$C_m \frac{\partial V_m}{\partial t} = -\frac{\partial}{\partial x}\left(D_{ex}\frac{\partial V_{ex}}{\partial x}\right) + I_{ion}(V_m, g_k), \tag{5}$$

$$C_m \frac{\partial V_m}{\partial t} = \frac{\partial}{\partial x}\left(D_{int}\frac{\partial V_{int}}{\partial x}\right) + I_{ion}(V_m, g_k), \tag{6}$$

where $V_m = V_{int} - V_{ex}$ is a transmembrane potential. Alternatively, a linear combination gives:

$$C_m \frac{\partial V_m}{\partial t} = -\frac{\partial}{\partial x}\left(D_{ex}\frac{\partial V_{ex}}{\partial x}\right) + I_{ion}(V_m, g_k), \tag{7}$$

$$\frac{\partial}{\partial x}\left(D_{ex}\frac{\partial V_{ex}}{\partial x} + D_{int}\frac{\partial V_{int}}{\partial x}\right) = 0. \tag{8}$$

These equations form the basis of bidomain models for cardiac tissue. In the one dimensional case, we can always reduce the number of equations from two to one as:

$$C_m \frac{\partial V_m}{\partial t} = \frac{\partial}{\partial x}\left(\frac{D_{ex}D_{int}}{D_{int}+D_{ex}}\frac{\partial V_m}{\partial x}\right) + I_{ion}(V_m, g_i). \tag{9}$$

For two or three dimensional media instead of a simple conductance we have a conductivity tensor, i.e. a matrix which gives the proportionality between voltage gradient and current density.

$$J_i = D_{ij}\frac{\partial V}{\partial x_j} \tag{10}$$

(here and throughout the rest of this section we assume the summation convention i.e. sum over doubly repeated indices) where V is potential, J_i is a vector of current density, and D is a symmetric tensor accounting for conductivity of the medium, which in two dimensions has the following form:

$$D = \begin{pmatrix} D_{11} & D_{12} \\ D_{12} & D_{22} \end{pmatrix}. \tag{11}$$

Following the same procedure as for one dimensional case we obtain:

$$C_m \frac{\partial V_m}{\partial t} = \frac{\partial}{\partial x_i}\left(D^{int}{}_{ij}\frac{\partial V_{int}}{\partial x_j}\right) + I_{ion}(V_m, g_k), \tag{12}$$

$$\frac{\partial}{\partial x_i}\left(D^{int}{}_{ij}\frac{\partial V_{int}}{\partial x_j} + D^{ex}{}_{ij}\frac{\partial V_{ex}}{\partial x_j}\right) = 0, \qquad (13)$$

where $D^{int}{}_{ij}$ and $D^{ex}{}_{ij}$ are conductivity tensors for intracellular and extracellular spaces.

We now want to eliminate re-entrant waves by applying external stimulation, and so need to obtain the response of cardiac tissue to externally applied stimulation.

3. Theory of single shock defibrillation

We begin with a bidomain approach (Biktashev et al., 1997), for a single isolated cell with an intracellular domain, $\mathcal{I}$, external domain, $\mathcal{E}$, and the membrane surface, $\mathcal{M}$, and introduce the electrostatic potential ϕ_i and ϕ_e in $\mathcal{I}$ and $\mathcal{E}$, u_i and u_e as limit values of ϕ_i and ϕ_e at $\mathcal{M}$, and electric charge densities q_i and q_e at the inside and outside surface of the membrane

$$\partial_t q_i = \Sigma_i \Delta_M u_i - f(u) + \sigma_i(\nabla \phi_i, \mathbf{n}), \qquad (14)$$

$$\partial_t q_e = \Sigma_e \Delta_M u_e + f(u) - \sigma_e(\nabla \phi_e, \mathbf{n}), \qquad (15)$$

here f, is the transmembrane current, $\Sigma_{i,e}$ are specific conductivities, and Δ_M is the Laplacian operator on the membrane surface. After defining $f(\mathbf{r},t), \mathbf{r} \in \mathcal{M}$, through local values of u, v and w, and local kinetic equations for v and w from the membrane excitation system, these equation form a closed system, which determine evolution of the distribution of electric properties over the cell at given $\mathbf{E}(t)$, and so describe the action of the external electric field onto the cell.

We will now simplify this extensive nonlinear system of partial differential equations to a system of equations of the form:

$$\begin{aligned} C\partial_t u &= f(u,v,w) + \hat{L}u + \hat{I}\mathbf{E} \\ \partial_t v &= g(u,v,w) \\ \partial_t w &= h(u,v,w) \end{aligned} \qquad (16)$$

where u, v and w are now functions of time and position on the membrane, $\hat{L}$ is a linear (generally, integro-differential) operator in a space of scalar functions on the membrane, and $\hat{I}$ is a linear operator mapping vectors $\mathbf{E}$ to scalar functions on the membrane. The specific forms of $\hat{L}$ and $\hat{I}$ depend on the geometry of $\mathcal{M}$ and on the coefficients $\sigma_{i,e}$ and $\Sigma_{i,e}$. We use the following properties of these operators:

- $\hat{L}u$ vanishes if u is spatially homogeneous over the membrane,

- the integral of $\hat{L}u$ over the membrane surface is zero for any u, and

- the integral of $\hat{I}\mathbf{E}$ over the membrane surface is zero for any $\mathbf{E}$.

We now construct a simplified model, that retains the main features of (16). We approximate all functions $u(x)$, $v(x)$ and $w(x)$ by piecewise constant functions, taking, at each time instant only two values at two different and fixed parts of the membrane. Denoting the two parts of the membrane by indices + and −, (16) is then rewritten in the form

$$\begin{aligned} C\partial_t u_+ &= f(u_+, v_+, w_+) + \alpha(u_- - u_+) + I_{ext}(t) \\ C\partial_t u_- &= f(u_-, v_-, w_+) + \alpha(u_+ - u_-) - I_{ext}(t) \\ \partial_t v_\pm &= g(u_\pm, v_\pm, w_\pm) \\ \partial_t w_\pm &= h(u_\pm, v_\pm, w_\pm) \end{aligned} \tag{17}$$

where the signs in the last two equations are either all + or −.

The constant α is the effective conductivity of the cell in this two-compartment approximation. $I_{ext}(t)$ is the current produced by the external source and crossing the cell.

The system (17) contains a singular small parameter, the ratio of the characteristic times of the intracellular conductivity α, τ_α, and of the membrane excitability, τ_f that can be excluded by "adiabatic" arguments; and if the duration of pulses I_{ext} is shorter than the characteristic time τ_h of the slow variables then $w_+ \approx w_- \approx W$. This final simplification gives:

$$\begin{aligned} C\partial_t U &= \tfrac{1}{2}(f(U + \tfrac{1}{2\alpha}I_{ext}(t), v_+, W) + f(U - \tfrac{1}{2\alpha}I_{ext}(t), v_-, W)) \\ \partial_t v_\pm &= g(U \pm \tfrac{1}{2\alpha}I_{ext}(t), v_\pm, W) \\ \partial_t W &= h(U, \tfrac{v_+ + v_-}{2}, W) \end{aligned} \tag{18}$$

This model is almost as simple as the ordinary differential equation for excitation of an isopotential cell - see equation (1) of my accompanying article - (e.g., it has three equations more than the 17 variable Oxsoft ordinary differential system we use for ventricular excitation), but describes the effect of external current. This has been obtained from (17) assuming the characteristic time of the external curent pulses, τ_I, is

$$0.1\text{ms} \sim \tau_\alpha \ll \tau_I \sim \tau_f \sim \tau_g \ll \tau_h \sim 10\text{ms}.$$

In practice τ_I, τ_f and τ_g are all of the order of 1 ms.

The general model (16) can be simplified by adiabatic arguements to

$$C\partial_t U = \int_M f(U + \hat{L}^{-1}\hat{I}\mathbf{E}(t), v, W)\, d\mathcal{M}, \tag{19}$$

with separate ordinary differential equations for v at each point of the membrane. If the external field $\mathbf{E}$ is fixed in direction and varies only in magnitude, then the surface integral in (19) can be reduced, in a Lesbegue style, to an ordinary integral

$$C\partial_t U = \int f(U + sE(t), v, W) K(s)\, ds, \qquad (20)$$

where the kernel $K(s)$ is determined by the cell geometry, the conductivities, and the direction of the external field, and because of electroneutrality of the cell

$$\int K(s)\, ds = 1.$$

The simple model (18) corresponds to evaluation of the integral in (20) at two points $s = \pm\frac{1}{2\alpha}$.

The strength-duration curve - the threshold external current I_{ext} as a function of stimulus duration for the two models (17) and (18) coincide with a very good precision - see Fig.2. Thus α is large enough for the adiabatic approximation to be valid.

The generalization of equations (18) for spatially extended tissue is straightforward as the deviation of each individual cell from an isopotential state takes place only during the short time periods of external stimulation, and outside these periods cable theory gives

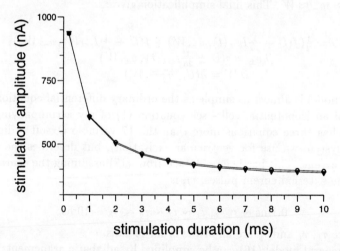

Figure 2: Excitation threshold, nA, as a function of stimulus duration, ms, in a single cell (open circles) and and simplified models (solid triangles) with $f()$, $g()$ and $h()$ were described by the Oxsoft guinea pig ventricle myocyte model that has 17 kinetic variables. α was $10\mu S$.

$$\partial_t U = \tfrac{1}{2C}(f(U + \tfrac{1}{2\alpha}I_{ext}(t), v_+, W) + f(U - \tfrac{1}{2\alpha}I_{ext}(t), v_-, W)) + D\partial_x^2 U$$
$$\partial_t v_\pm = g(U \pm \tfrac{1}{2\alpha}I_{ext}(t), v_\pm, W)$$
$$\partial_t W = h(U, \tfrac{v_+ + v_-}{2}, W),$$
(21)

where U, $v_\pm$ and W are now functions not only of time t, but also of distance along the fibre x, and the diffusion coefficient for voltage, D, is proportional to the intercellular conductivity. $D = 1.25 \text{cm}^2/\text{s}$ gives a conduction velocity of $0.6 \,\text{m/s}$ for a solitary wave through resting tissue. The value of D is necessary only for the interpretation of spatial scales as equations (21) are invariant under simultaneous change of spatial scales and coefficient D.

We now apply this approach to evaluate the defibrillation threshold for a tissue, i.e. the amplitude of an externally applied current pulse necessary to abolish all propagating waves, and compare it with the prediction of the asymptotic theory of defibrillation for our model - see Fig. 3. This theory was described in Keener (1996) and Pumir & Krinsky (1996) and is based on separate consideration of the slow and fast processes during the process of propagation (Fife, 1976; Tyson & Keener, 1988), and assumes that on the fast

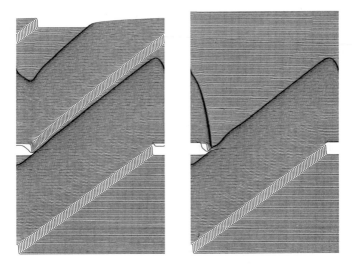

Figure 3: (a) Subthreshold (600nA, 2 ms) and (b) suprathreshold (800nA, 2 ms) response of a re-entrant wave in one-dimensional ring to a defibribrillating shock. In both cases the wavefront is initially advanced, in the suprathreshold case the wavefront collapses back to its position at the time the pulse was applied, and meets its waveback.

time scale the medium has two alternative stable equilibria, which depend on the slow variables. The propagation of the wavefront, in the fast time scale, is a trigger wave between the equilibria that is either "antegrade", when the excited region grows, or "retrograde" propagation, when the excited region shrinks. The wavefront of a propagating pulse is an antegrade trigger wave, and its back a retrograde wave. In a resting medium, the upper, "excited" equilibrium is more stable, so a suprathreshold perturbation produces an antegrade trigger wave, and the excited region expands. During the excited state, the evolution of the slow variables lowers the stability of the excited state while the stability of the resting state increases, until a retrograde trigger wave can propagate.

The boundary between these classes is in the state space of the slow variables, and corresponds to the values of these variables when the two equilibria are equally stable. If u is much faster than both v and w, this "equal stability" is represented by a "Maxwell rule", for the right-hand side of the fast excitability equation:

$$\int_{u_{resting}(v,w)}^{u_{excited}(v,w)} f(u,v,w)du = 0. \tag{22}$$

In biophysical excitation equations and so the theory is applicable as long as the "slow processes" v and w are slow enough for the asymptotic approach to be valid. The margin (22) between preferably-excited and preferably-resting cells is a manifold of codimension 1, and so finding the exact defibrillation condition would imply finding the points with the highest threshold on this manifold. We do not need to search the whole space, only in its subset, corresponding to states of cells present in the tissue in the moment of defibrillation. For a spiral wave solution of the two-dimensional analogue of (21), there is a narrow gap between the waveback of the spiral wave and the following wavefront. The response of such a counter-clockwise rotating re-entrant spiral (Fig.4) to a brief defibrillating pulse is similar to that seen in the one-dimensional model: the wavefront is forced forward, and then relaxes back to its position at the time the shock was applied, while the waveback continues to rotate counter-clockwise. All activity is extinguished when the waveback reaches the wavefront.

4. Defibrillation by resonant drift

A spiral wave can be forced to move by a spatially uniform, time periodic perturbation of appropriate frequency (Agladze, 1987; Davydov, 1988) . We have presented a phenomenological theory for such resonant drift (Biktashev & Holden, 1993). Resonant drift in the location of a spiral occurs when the frequency of perturbation is the same as the frequency of rotation of the spiral; in effect, each perturbation is applied at the same phase of rotation and so has the same effect on the spiral; since the spiral is stable this effect can only be a displacement in tip position and/or a change in phase of the spiral. This general property of excitable media has been exploited as a means of moving spiral waves to the boundaries of the medium. A boundary can repel a resonantly drifting spiral, and this reflection of the spiral can be overcome by appropriate timing of the repetitive perturbation, and so the spiral can be driven onto the inexcitable boundaries and be extinguished (Biktashev & Holden, 1993). In principle, resonant drift under feedback control could provide a means of eliminating re-entrant activity in cardiac tissue by using small amplitude, repetitive electrical stimulation (Biktashev & Holden, 1994). This will be practical only if any re-entry is eliminated within a reasonable time, say less than 30s.

We have used resonant drift to move and control re-entrant waves in simplified FitzHugh-Nagumo models of excitable media, and we have shown that using feed-back control of the stimulation can provide elimination of the re-entrant activity, even when the period of re-entry is unknown or is varying in time, or in presence of inhomogeneities, or when there are multiple re-entrant sources

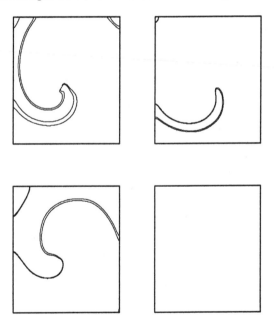

Figure 4: Wavefront and waveback of a spiral wave solution before application of the defibrillating pulse, 20 ms , 80 ms and 120 ms after application of the defibrillation pulse, when all activity is extinguished.

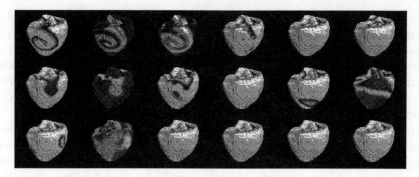

Figure 5: Frames from a movie illustrating elimination of re-entrant activity in canine ventricle CML model by spatially uniform, low amplitude stimulation triggered by the arrival of a wavefront at a point on the right epicardial surface. The first few stimuli leave re-entrant sources deep in the heart, so activity re-emerge on the surface. However, finally (within a few s of canine time) all re-entrant activity is eliminated.

present (Biktashev & Holden, 1993,1994,1995a). Defibrillation by resonant drift under feedback control for the ventricle CML model is illustrated in Fig. 5.

These computations have been extended to 2-D atrial and ventricular tissue models (Biktashev & Holden, 1995b, 1996). The effect of strictly periodic repetitive stimulation is to produce a drift of the position of the spiral, and general theory (Davydov et al., 1988; Biktashev & Holden, 1995a) predicts that this drift would be along a circle, with a drift velocity that depends upon stimulation amplitude and an angular velocity equal to the difference between the stimulation frequency and the resonant frequency of the spiral, where the latter also depends on stimulation amplitude and is close to the frequency of free spiral rotation when the amplitude of stimulation is small. Such an induced circular drift is illustrated in Fig. 6, so the behaviour of the atrial tissue model is still consistent with the general theory, even though the theory was developed for a medium without refractoriness (Davydov et al., 1988), and for a medium with a refractory period but using linear perturbation theory (Biktashev & Holden, 1995a) which is not formally applicable to the atrial tissue model because of the steepness of the propagation front.

Strictly periodic stimulation is not very effective in extinguishing the spiral, as it is difficult to choose appropriate conditions that would cause the spiral wave to reach the inexcitable boundaries. Drift in a straight line towards the boundary is produced only in the case of an exact resonance, and even in this case the drift may not cause annihilation of the spiral, as the resonantly drifting spiral can be reflected by the boundary (Biktashev & Holden, 1993).

205

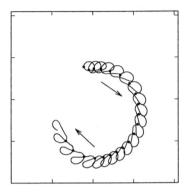

Figure 6: Effect of strictly periodic repetitive stimulation with period $77.2ms$ and amplitude $4mV/ms$ on the trajectory of the tip of the spiral (continuous curve) for the rabbit atrial tissue model, and the phase (•) of the spiral, given by the position of the tip at instants of each next stimulus. The averaged trajectory of the tip is nearly circular. The angular velocity of the averaged drift is the difference between the vortex frequency and the stimulation frequency.

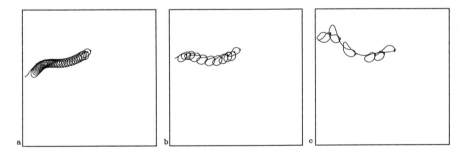

Figure 7: Tip trajectories under feedback controlled, resonant driving of atrial tissue model. Stimulation amplitude doubles between each panel.

However, feed-back controlled stimulation is more effective at driving the spiral to the boundaries (Biktashev & Holden, 1994). The feed-back dynamically adjusts the frequency of stimulation to the instantaneous frequency of rotation, thus damping the de-tuning of resonance caused by changes in the frequency of rotation as the resonantly drifting spiral approaches a boundary, that leads to reflection of resonantly drifting spirals.

Repetitive stimulation, triggered each time the wavefront reaches a recording site produces a drift of the spiral, which in the standard atrial tisue model always drives the spiral onto the boundary (Fig. 7). In FitzHugh-Nagumo models, the interaction between the resonantly drifting spiral and the boundary was strong enough and the drifting spiral could stop near the boundary and persist, if the simulation amplitude was below a certain threshold. In the atrial tissue model, the interaction between the spiral and the boundary is weak and the annihilation was always produced, even at lowest stimulation amplitudes used ($0.5mV/ms$).

In the ventricular tissue model, that shows a complicated pattern of meander, changing from a linear to a multi-lobed core, resonant drift can still be use to drive the meandering spiral onto the medium boundaries (Fig. 8) (Biktashev & Holden, 1996).

The repetitive stimuli used in the above computations are all less than the threshold for a single "defibrillating shock" of the same duration, i.e. the

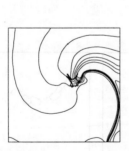 

Figure 8: *left* Contours of re-entrant wave moving around an irregular, multi-lobed core and *right* tip trajectories under feedback controlled, resonant driving of ventricular tissue model. When the wavefront of the spiral wave (depolarisation through -10mV) reached a recording site in the bottom left hand corner, a 2ms, 4V/s depolarising perturbation was added after a fixed delay. Each trajectory is for a different delay, from 0 to 100ms, and corresponds to applying the perturbation at a different phase of the spiral.

minimal shock which is sufficient for suppressing the spiral.

The surprising results of these computations are that, in spite of the strong nonlinearities of the excitation equations, the main qualitative predictions of the phenomenological theory for feed-back driven resonant drift developed in Biktashev & Holden (1994,1995a) have proved to be valid (except for linear dependence upon stimulation amplitude), and that even with stimulus sizes 10% of the single shock defibrillation threshold drift velocities of $1 cms^{-1}$ are reached. Despite all the simplifications made, at least the order of magnitude of this velocity should be realistic. Thus we anticipate that resonant drift produced by low amplitude, spatially uniform repetitive stimulation under feedback control can extinguish re-entrant arrhythmias in both mammalian atrial and ventricular tissue within about $10 - 30s$, and so can provide a practical technique for extinguishing flutter and monomorphic tachycardia.

6. Current problems

In spite of its practical importance, the processes of defibrillation remain obscure. Most theoretical approaches have been based on linear models (Knisley, et al., 1994; Sepulveda, et al., 1989; Krassowska et al., 1990) while some numerics of nonlinear models (Cartee & Plonsey, 1992) and theoretical studies with simplified models have been attempted (Pumir and Krinsky, 1996). The reason for this lack of progress is the combination of nonlinearity with the necessity for a representation of the complicated spatial structure for every cell. We have overcome these problems by applying a series of well known methods, to fullfill nonlinear averaging; the result is the simplified models of (17)and (21). This reduction of an infinite dimensional system to an ordinary differential system may be of value in a range of applications of nonlinear science.

These simplified models have been verified numerically and allow us to use biophysically detailed excitation equations, and so we are now in a position to provide a quantitative, theoretical explanation for the effects of changes in parameters in the excitation equations on the defibrillation threshold, and to design optimal defibrillation pulse parameters. Experimental techniques now exist (Zhou et al., 1995) for testing such quantitative descriptions of the mechanisms of defibrillation.

Acknowledgement

The work described here was done in collaboration with Dr. Biktashev and supported by grants from the Wellcome Trust (042352, 044365) and the EPSRC ANM initiative (GR/J35641).

References

Agladze, K.I., V.A.Davydov, & A.S.Mikhailov, (1987). " An observation of resonance of spiral waves in distributed excitable medium.", *Pis'ma v ZETP* **45**, 601-603 (in Russian, see English translation in *Sov.Phys.-JETP Letters*).

Biktashev, V.N. & A.V.Holden, (1993). "Resonant drift of an autowave vortex in a bounded medium", *Physics Lett. A* **181**, 216-224.

Biktashev, V.N. & A.V.Holden, (1994). "Design principles for a low voltage cardiac defibrillator based on the effect of feedback resonant drift", *J. Theor. Biol.* **169**, 101-112.

Biktashev, V.N. & A.V.Holden, (1995a). " Resonant drift of autowave vortices in 2D and the effects of boundaries and inhomogeneities", *Chaos Solitons and Fractals* **5**, 575-622.

Biktashev, V.N. & A.V.Holden, (1995b). " Control of re-entrant activity in a model of mammalian atrial tissue", *Proc. Roy. Soc. Lond. B* **260**, 211-217.

Biktashev, V.N. & A.V.Holden, (1996). " Re-entrant activity and its control by resonant drift in a two-dimensional model of isotropic homogeneous ventricular tissue" *Proc. Roy. Soc. B* **263** 1373-1383.

Biktashev, V.N. , A.V.Holden, & H.Zhang, (1997). "A model for the action of external current onto excitable tissue", *International J of Bifurcation and Chaos* **7** (2) in press.

Davydov, V.A., V.S.Zykov, A.S.Mikhailov, & P.K.Brazhnik, (1988). "Drift and resonance of spiral waves in distributed media", *Izv. Vuzov - Radiofizika* **31**, 574-582 (in Russian, see English translation in *Sov.Phys.-Radiophysics*).

Fife, P.C. (1976). " Boundary and interior transition layer phenomena for pairs of second-order differential equations", *J.Math.Anal.Appl.* **54**, 497 -521.

Holden, A. V., M.J.Poole, & J.V.Tucker, (1996). " An algorithmic model of the mammalian heart: propagation, vulnerability, re-entry and fibrillation", *International J. of Bifurcation and Chaos* **6** 1623-1635.

Keener, J.P. (1996). " Direct activation and defibrillation of cardiac tissue", *J. theoretical biology* **178** 313-324.

Knisley, S.B., B.Hill, & R.E.Ideker (1994)." Virtual electrode effects in myocardial fibre", *Biophys. J.* **66**, 719-728 .

Krassowska, W., D.W.Frazier, T.C.Pilkington, & R.E.Ideker, (1990). "Potential distribution in three-dimensional anisotropic medium", *IEEE Biomed. Eng.* **37**, 267-284.

Panfilov, A.V. & A.V.Holden, (eds) (1997) . *The computational biology of the heart* . J.Wiley, Chichester.

Pumir, A. & V.Krinsky,. (1996). "How does an electric current defibrillate cardiac muscle", *Physica D*, **91** 205-219.

Sepulveda, N.G, B.J.Roth, & J.P.Wikswo Jr , (1989). "Current injection into a two-dimensional anisotropic medium", *Biophys J* **55** , 987-999.

Tyson, J. J. & J.P.Keener, (1998). "Singular perturbation theory of traveling waves in excitable media (a review)", *Physica D* **32**,327-361.

Zipes, D.P. & J.Jalife, (eds) (1995). *Cardiac Electrophysiology - from cell to bedside*. W.B.Saunders, Philadelphia.

Zhou,X., R.E.Ideker, T.F.Blitchington, W.M.Smith, & S.B.Knisley, (1995). "Optical transmembrane potential measurements during defibrillation-strength shocks in perfused rabbit hearts",*Circulation Research* **77**, 593-602.

NOISE AND STOCHASTIC RESONANCE

CHAOS AND NOISE IN EXCITABLE SYSTEMS

ROLANDO CASTRO and TIM SAUER
Institute of Computational Sciences and Informatics
and Department of Mathematical Sciences
George Mason University
Fairfax, VA 22030, USA
tsauer@gmu.edu

ABSTRACT

In principle, the state space of a chaotic attractor can be partially or wholly reconstructed from interspike intervals recorded from experiment. Under certain conditions, the quality of a partial reconstruction, as measured by the spike train prediction error, can be increased by adding noise to the spike creation process. This phenomenon for chaotic systems is an analogue of stochastic resonance.

1. Introduction

Attractor reconstruction, summarized in (Castro & Sauer, this volume), can be viewed as a type of information transfer. In the case of a topological embedding, no information is lost. The set of states of the underlying experimental system is reproduced exactly in the copy that is reconstructed from measured data. In other cases, the reconstruction may be incomplete. When a chaotic signal is fed into a threshold crossing detector, the time-delay plot of time intervals between crossings reconstructs something akin to a Poincaré section of the underlying chaotic attractor. The dimension is decreased by one (Castro & Sauer, 1997). A recent study (Racicot & Longtin, 1997) of neuron models subjected to chaotic input points out that the two-dimensional FitzHugh-Nagumo differential equation (FitzHugh, 1961; Nagumo et al., 1962) (hereafter referred to as FHN2) acts as a kind of threshold-crossing "filter" for input signals, and in particular fails to completely reconstruct the attractor which generated the input signal. One focus of the present article is to clarify the distinction, for attractor reconstruction purposes, between threshold-crossing (TC) and integrate-and-fire (IF) filters. In contrast to the fact that FHN2 acts as a TC filter, we exhibit a variation on FHN2 which acts principally as an IF filter instead, and does fully reconstruct the input attractor. This differential equation model, which we shall denote FHN3, is a three-dimensional version of excitable FitzHugh-Nagumo-type dynamics. With appropriate parameter settings, it has the following remarkable property: When a signal from a system attractor is added to one of FHN3's variables, another variable undergoes a deterministic spiking behavior whose interspike intervals, embedded as m-tuples, embed the original attractor in m-dimensional reconstruction space. Just as Takens' theorem guarantees that multidimensional state space information can be condensed into a single evenly-spaced time series, the same can be accomplished with spike timings from the FHN3 filter.

Viewing attractor reconstruction as information transfer raises questions about the efficiency of the transfer process, and the possible effects of noise on this process. Surprisingly, in some instances noise can have a beneficial effect on information transmission, analogous to the stochastic resonance phenomenon (Weisenfeld & Moss, 1995; Moss et al., 1993) observed for multistable potentials and excitable media. This seemingly contradictory effect is the observed amplification of a filtered signal achieved when stochastic noise is added to the input signal.

Stochastic resonance has been shown to amplify signals generated by linear models, such as sine waves. Recently, it has been shown (Collins et al., 1995) that the same basic effect can be seen with aperiodic (stochastic) input signals if the means to measurement is appropriately modified. Since the spectrum of a random signal is not discrete, the SNR must be replaced in this case with a power norm sensitive to shape-matching and/or signal correlation.

Our present interest in stochastic resonance is somewhat different. Our goal is in maximizing the amount of state information carried by the spike train. We pose the question whether the quality of state information from a *deterministic* signal (as opposed to the stochastic realization used in (Collins et al., 1995)) can be improved by injecting random noise into the process. In particular, we are interested in the case where the input signal is chaotic. To determine the quality of state information contained in the spike train, we measure the ability to predict the spike train from its own history, using nonlinear prediction techniques. We will show examples in which the interspike interval prediction error of the output signal decreases (predictability increases) with increasing noise added to the input signal. Showing that nonlinear predictability is enhanced by adding noise is analogous to the enhanced SNR shown in stochastic resonance studies.

2. Spike-generating filters

The filters we will use to create spike trains capable of carrying low-dimensional deterministic state information are based on the well-known FitzHugh-Nagumo equation. The two dimensional system FHN2

$$\epsilon \dot{v} = -v(v - 0.5)(v - 1) - w + S$$
$$\dot{w} = v - w - b \tag{1}$$

is a simple differential equation that exhibits a fast spike followed by a refractory period. There is an equilibrium at $(v, w) = (v_0, v_0 - b)$, where v_0 is a real-valued root of $v_0(v_0 - 0.5)(v_0 - 1) = S + b - v_0$. A stability check of the equilibrium v_0 shows the existence of a supercritical Hopf bifurcation for $S_H = v_H(v_H - 0.5)(v_H - 1) + v_H - b$, where $v_H = 0.5 - \sqrt{3 - 12\epsilon}/6$. Therefore, if b, ϵ are fixed and the bifurcation parameter S is increased, the system undergoes a Hopf bifurcation at S_H, resulting in a periodic orbit of the system encircling the formerly stable equilibrium. The periodic orbit is manifested in rhythmical spiking by the variable v. For example, setting $b = 0.15, \epsilon = 0.005$, there is a Hopf bifurcation point at $S_H \approx 0.112331\ldots$. For $S < S_H$, the system is quiescent; the equilibrium is stable. For $S > S_H$, the system spikes at a rate of approximately 1 Hz.

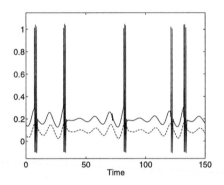

Figure 1: The solid curve is the variable v of FHN2, the FitzHugh-Nagumo equation (1) with s replaced by $S(t) = 0.09 + 0.013x(t)$, where $x(t)$ is a solution of the Rössler system (2). The dashed curve is $S(t)$. When a peak of $S(t)$ is greater than ≈ 0.15, a burst is triggered in FHN2.

Now consider FHN2 as a nonlinear filter by substituting for the constant S in (1) a signal $S(t)$ from another system. Fig. 1 shows a plot of the variable v from (1) where S has been replaced by a signal from the (Rössler, 1976) system

$$\begin{aligned} \dot{x} &= \tau(-y - z) \\ \dot{y} &= \tau(x + ay) \\ \dot{z} &= \tau(b + (x - c)z) \end{aligned} \quad (2)$$

where the standard parameters are set to $a = 0.36, b = 0.4, c = 4.5$, and $\tau = 0.5$ causes the trajectory to run at half speed. The input signal, also plotted in Fig. 1, is $S(t) = 0.09 + 0.013x(t)$, where $x(t)$ is the x variable of (2). The bias of the signal is $\langle S(t) \rangle = 0.093$, and its root mean square amplitude is $\sqrt{\langle (S(t) - \langle S(t) \rangle)^2 \rangle} = 0.035$. Fig. 2 demonstrates the threshold-crossing detection capability of FHN2. When the peak height of $S(t)$ is greater than ≈ 0.15088, FHN2 fires a burst of spikes. Note that this threshold is significantly higher than the Hopf bifurcation value s_H, which would be the threshold in the limit of an $S(t)$ which oscillates infinitely slowly.

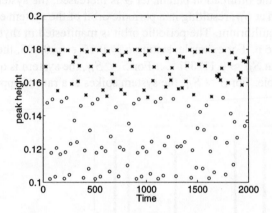

Figure 2: Peak heights of the signal $S(t)$ from Fig. 1 graphed versus time. The height is plotted as an asterisk if it triggers a burst from FHN2; as an open circle if not. All of the asterisks lie below all of the open circles, signifying precise threshold detection by FHN2.

3. Noise and predictability

Although we expect the spike sequences to carry state information of the Rössler system, because of its threshold detection behavior we do not expect it to carry enough to reconstruct the entire attractor. On the other hand, the benefit of a TC filter is that noise can in some cases enhance the reconstruction quality, as measured by prediction error. As in studies of stochastic resonance, we will add white noise to the input signal of the filter (in this case, the FHN2 spike generator). The equation with noise term is

$$\epsilon \dot{v} = -v(v - 0.5)(v - 1) - w + S(t) + \xi(t)$$
$$\dot{w} = v - w - b, \qquad (3)$$

where $\epsilon = 0.005$, $b = 0.15$, and $\xi(t)$ is Gaussian white noise with zero mean and autocorrelation $\langle \xi(t)\xi(s) \rangle = 2D\delta(t-s)$. For small values of the noise level D (including all those considered here), the variable v exhibits a clearly distinguishable spiking behavior, often in bursts of more than one spike, as in Fig. 1. For analysis purposes, we found it more convenient to collect series of interburst intervals, each defined to be the elapsed time between the final spike of one burst and the first spike of the next burst. After using (3) to make a series of 1024 interburst intervals, we used a standard nonlinear prediction algorithm to measure the level of determinism in the series. The fact that state information from a deterministic system is contained in a spike train, even when the spike train is chaotic, can be detected by measuring the nonlinear predictability of the interburst intervals. If it can be shown that the ISI series is predictable "beyond the power spectrum", that is, if there is predictability beyond that which is guaranteed by linear autocorrelation, then there is evidence of nonlinear dynamics in the series.

The prediction algorithm works as follows. Let $V_0 = (t_{i_0}, \ldots, t_{i_0-m+1})$ be an ISI vector. The 1% of other reconstructed vectors V_k that are nearest to V_0 are collected, omitting vectors V_k close in time. The ISI for some number h of steps ahead are averaged for all k to make a prediction. That is, the average $p_{i_0} = \langle t_{i_k+h} \rangle_k$ is used to approximate the future interval t_{i_0+h}. The difference $p - t_{i_0+h}$ is the h-step prediction error at step i_0. We could instead use the series mean m to predict at each step; this h-step prediction error is $m - t_{i_0+h}$. The ratio of the root mean square errors of the two possibilities (the nonlinear prediction algorithm and the constant prediction of the mean) gives the normalized prediction error NPE = $\langle (p_{i_0} - t_{i_0+h})^2 \rangle^{1/2} / \langle (m - t_{i_0+h})^2 \rangle^{1/2}$ where the averages are taken over the entire series. The normalized prediction error is a measure of the (out-of-sample) predictability of the ISI series. A value of NPE less than 1 means that there is linear or nonlinear predictability in the series beyond the baseline prediction of the series mean.

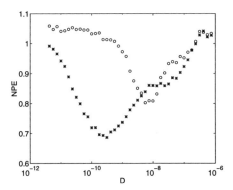

Figure 3: Normalized prediction error of spike trains generated by (3), where $S(t)$ is a signal formed using the Rössler x-variable of (2) with bias 0.075 and rms amplitude 0.020 (open circles) or 0.023 (asterisks). As the input noise power D increases, the NPE displays a minimum, corresponding to maximum information transfer. Each plotted point is an average over 12 noise realizations; standard error is less than 0.02 for each.

The results of the predictability of the interburst interval series from (3) are shown in Fig. 3. For these parameter settings, unlike those for Fig. 1, no spikes occurs in the absence of noise. As the noise power D is increased from zero, spikes begin to occur for very small noise levels, although the interburst series show no predictability (NPE $\approx$ 1) until D is raised beyond 10^{-11}. The prediction error then drops to a minimum and raises again when the noise becomes large enough to swamp the system. The clearly noticeable improvement in predictability due to extremely small noise input is essentially a stochastic resonance effect. These results show evidence of nonlinear determinism, since Gaussian-scaled surrogate series (Theiler et al., 1992) created from all burst series considered in Fig. 3 have NPE $\approx$ 1.

4. A faithful FitzHugh-Nagumo filter

A slight alteration in the FitzHugh-Nagumo equations yields a nonlinear filter that acts as an integrate-and-fire processor. Define the system FHN3 by

$$
\begin{aligned}
\dot{u} &= -au - cw + S(t) \\
\epsilon \dot{v} &= -v(v - 0.5)(v - 1) + u - dw \\
\dot{w} &= v^2 - w - b.
\end{aligned}
\tag{4}
$$

This system is similar to FHN2 in that if $S(t)$ is set to be a constant parameter S, there is a Hopf bifurcation as S is increased. Setting parameters $a = 0.1, b = 0.15, c = 0.5, d = 0.5, \epsilon = 0.005$, the bifurcation point is $S_H \approx -0.059$.

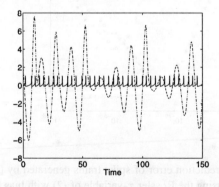

Figure 4: The solid curve is the variable v of FHN3, Eq. (4), with $S(t) = 0.0023x(t) - 0.04$. The dashed curve is $x(t)$, the x-variable of the Rössler attractor (2). A plot of 3-tuples of interspike intervals from this equation is shown in Fig. 5a.

Its success as an information processor is shown in Fig. 4. As with FHN2, we replace the parameter S with an input signal $S(t)$ from the Rössler attractor. The signal is $S(t) = 0.0023x(t) - 0.04$, which corresponds to bias $\langle S(t) \rangle = -0.04$ and root mean square amplitude is $\sqrt{\langle (S(t) - \langle S(t) \rangle)^2 \rangle} = 0.006$. Comparing with Fig. 1, we see a marked difference in the way FHN3 processes the input signal compared to FHN2.

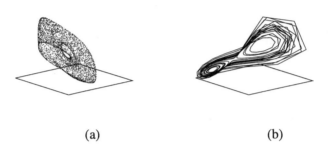

(a)   (b)

Figure 5: Interspike interval reconstructions of (a) the Rössler-FHN3 intervals from Fig. 4 (b) Lorenz-FHN3 intervals from (4) with $S(t) = .0005x(t) - 0.04$, where $x(t)$ is the x-variable of the Lorenz system. In (b), fewer points are plotted, and they are connected with line segments.

Fig. 5(a) shows a three-dimensional plot of the vectors (t_i, t_{i-1}, t_{i-2}), where $t_i = T_i - T_{i-1}$ is the time interval between spikes of the v variable of FHN3. Fig. 5(b) shows a reconstruction where the input signal $S(t)$ is the x-coordinate from the (Lorenz, 1963) equations $\dot{x} = \tau(\alpha(y - x)), \dot{y} = \tau(\rho x - y - xz), \dot{z} = \tau(-\beta z + xz)$, where the parameters are set to the standard values $\alpha = 10, \rho = 28, \beta = 8/3$, and $\tau = 0.01$. Apparently, the interspike intervals recovered from (4) do an effective job of reconstructing the chaotic attractor which produced the input signal $S(t)$, for both the Rössler and Lorenz examples. Nonlinear prediction on a length 1024 series of spikes created as in Fig. 4 yield NPE $= 0.1$. This very low NPE supports the visual indication in Fig. 5(a) of a faithful reconstruction of the underlying Rössler attractor. This is similar to the mechanism that was studied in the generic integrate-and-fire model of (Sauer, 1994), where firing times T_i were generated recursively by

$$\int_{T_i}^{T_{i+1}} S(t)dt = \Theta \tag{5}$$

for a fixed threshold Θ. Theoretical reconstruction results for spike trains generated by model (5) are discussed in (Sauer, 1996).

5. Conclusions

Creating spikes using FHN2 or FHN3 means imposing a type of highly nonlinear filter on the attractor signal, a filter which edits out amplitude information (since the spike waveforms are essentially alike) and converts the information entirely to event timings. Our purpose is to gain insight into the data processing methods used in systems which communicate through spike timings, as is conjectured for certain neural systems (Abbott, 1994; Bialek et al., 1991). We have shown by example that noise may be useful for this communication, in that it can amplify transmission of deterministic, nonlinear state information as measured by nonlinear prediction error. For the latter spike generation model (FHN3), we have the possibility of complete reconstruction of attractor states.

Acknowledgements

The research of T.S. was supported in part by the U.S. National Science Foundation (Computational Mathematics program).

References

Abbott, L. (1994), *Quart. Rev. Biophys.* 27:291.

Bialek, W., F. Rieke, R. R. de Ruyter van Steveninck, D. Warland (1991), *Science* 252:1854.

Castro, R., T. Sauer (1997) "Correlation dimension of attractors through interspike intervals", *Phys. Rev. E* 55:287-290.

Castro, R., T. Sauer (1998) "System recontruction and analysis using interspike intervals", this volume.

Collins, J.J., C.C. Chow, T.T. Imhoff (1995) *Phys. Rev. E* 52:R3321.

Fitzhugh, R. (1961) *Biophys. J.* 1:445.

Lorenz. E. (1963) "Deterministic nonperiodic flow", *J. Atmos. Sci.* 20:131.

Mainen, Z., T. J. Sejnowski (1995) *Science* 268:1503.

Moss, F., A. Bulsara, M.F. Schlesinger (1993) Eds. Proc. of NATO Advanced Research Workshop on Stochastic Resonance in Physics and Biology, *J. Stat. Phys.* 70.

Nagumo, J., S. Arimoto, S. Yoshizawa (1962) *Proc. IRE* 50:2061.

Racicot, D., A. Longtin (1997) "Interspike interval attractors from chaotically driven neuron models", *Physica D*.

Rössler, O. E. (1976) "An equation for continuous chaos", *Phys. Lett.* 57A:397.

Roux, J. C., A. Rossi, S. Bachelart, C. Vidal (1980) "Representation of a strange attractor from an experimental study of chemical turbulence", *Phys. Lett.* A77:391.

Sauer, T. (1994) "Reconstruction of dynamical systems from interspike intervals", *Phys. Rev. Lett.* 72:3911.

Sauer, T. (1996) "Reconstruction of integrate-and-Fire dynamics", in *Nonlinear Dynamics and Time Series,* eds. C. Cutler, D. Kaplan. Fields Institute Publications, Amer. Math. Soc., Providence, RI.

Theiler, J., S. Eubank, A. Longtin, B. Galdrakian, and J.D. Farmer (1992) "Testing for nonlinearity in time series: The method of surrogate data", *Physica D* (Amsterdam) 58:77.

Vaadia, E., I. Haalman, M. Abeles, H. Bergman, Y. Prut, H. Slovin, A.M.H.J. Aertsen (1995) *Nature* 373:5151.

Weisenfeld, K., F. Moss (1995) *Nature* 373:33.

COMMUNICATIONS

DETECTION OF SMALL-CONDUCTANCE Cl⁻ CHANNELS BY WAVELET TRANSFORM

M. NOBILE, L. LAGOSTENA and R. FIORAVANTI

Istituto di Cibernetica e Biofisica, C.N.R., Via De Marini 6, 16149 Genova, Italy

E-mail: nobile@icb.ge.cnr.it

and

S. FERRONI

Dipartimento di Fisiologia Umana e Generale, Universita' di Bologna,
Via San Donato 19/2, 40127 Bologna, Italy

ABSTRACT

The biophysical and pharmacological properties of an inwardly rectifying Cl⁻ conductance expressed in cultured rat neocortical astrocytes have been recently investigated by using the whole-cell patch-clamp technique. In this study we report on the detection of single Cl⁻ channel activity in outside-out patches by means of Wavelet functions. Power spectra clearly showed, in the low frequency domain (up to 20 Hz), a significant difference between current traces obtained under control conditions at -100 mV and those recorded at the same potential in the presence of zinc, a blocker of the inwardly rectifying Cl⁻ conductance, or with traces at 0 mV where the single channel current is strongly reduced. The current traces used for the power spectra were filtered by means of Wavelet functions. Since all the evidences were in good agreement with the hypothesis of recordings containing single channel events, we decided to use the Haar decomposition. At -100 mV in the presence of zinc or at 0 mV under control conditions, the filtered current signal showed small symmetrical transitions as expected for white- noise signal. Conversely, in the control traces at -100 mV, the filtered signal clearly indicated the presence of negative, abrupt transitions in the low frequency domain which resembled the gating of the single Cl⁻ channel. These results demonstrate that the macroscopic inwardly rectifying Cl⁻ conductance expressed in rat astrocytes is due to the activity of a low-conductance single channels, and that the wavelet transform can be considered to be a new approach to the study of ion channels.

Introduction

While for high conductance Cl⁻ channels the elementary current properties are well known and characterised, the properties of low conductance Cl⁻ channels are still poorly understood despite their widespread occurrence and essential physiological role (Woll et al., 1987; Pusch et al., 1994). This is due to the small unitary current that render them not clearly distinguishable from noise in conventional single-channel patch-clamp measurements.

In the present study we utilised stationary noise analysis and Wavelet functions to determine the unitary properties of the inwardly rectifying Cl⁻ conductance expressed in cultured rat neocortical astrocytes following a long-term treatment (1-3 weeks) with dibutyryl-cAMP (Ferroni et al., 1995; Ferroni et al., 1997). Analyses were performed on outside-out single channel currents. The results demonstrate that by means of Wavelet transform it is possible to detect the activity of low-conductance single Cl⁻ channels.

Materials and Methods

Patch-clamp experiments were performed on cultured type-1 cortical astrocytes prepared from 1-2-day-old rats, as previously described (Ferroni et al., 1997). The composition of the standard external solution was (mM): 140 N-methyl-D-glucamine chloride (NMDG-Cl), 4 KCl, 2 $CaCl_2$, 2 $MgCl_2$, 5 N-tris-hydroxymethyl-2-aminoethane sulphonic acid (TES), 5 glucose. The standard internal solution was (mM): 144 NMDG-Cl, 2 $MgCl_2$, 5 TES, 5 EGTA. The pH were adjusted to 7.3 with $NMDG^+$ and the osmolarity were set to 330 mOsmol with mannitol. Unitary channel currents were recorded using the patch-clamp technique in the outside-out configuration. Borosylicate glass electrodes were pulled and calibrated to have a tip resistance of ≈10 MΩ when filled with the standard internal solution. Current records, 10 s long, were obtained at membrane potentials between 0 to -120 mV, filtered at 10 kHz, digitised with a sampling time of 30 μs and were not corrected for leakage. The power spectra in all the experimental situations have been computed using FFT (Fast Fourier Transform) over a sample of $N=2^{16}$ data in order to detect the global difference between currents recorded at -100 mV membrane potential in control conditions and in the presence of external 1 mM zinc, both, or currents elicited from membrane potentials of 0 mV. Since in our case the noise is predominant to the signal, we used a simple and powerful procedure. The detection of elementary currents small in amplitude and stochastic in duration, suggested the use of Haar Wavelet function, $\Psi(t)=1$, $0<t<0.5$; $\Psi(t)=-1$, $0.5 \leq t<1$, for a dyadic signal. The unit norm function

$$\Psi_\lambda(t)= 2^{j/2} \Psi(2^j t-k) \quad (1)$$

with j = dilatations, k = shifting; for each λ (j,k) (j and k integers) is such that $(\Psi_\lambda)_\lambda$ is a complete orthonormal system. Hence, if we define the wavelet coefficients

$$\alpha_\lambda = \int \Psi_\lambda(t)\, f(t)\, dt \quad (2)$$

we have the reconstruction formula

$$f = \sum_\lambda \alpha_\lambda \Psi_\lambda \quad (3)$$

In our case denoising is to keep in (3) the $\lambda_i \in \lambda$ approximating the channel opening and closing. We used the MatLab-Wavelet Toolbox package (The Math Works Inc., Natick, Mass.). The type of signal suggested not to use the simple Wavelet decomposition tree in approximation and detail, but the complete tree decomposition of Wavelet-Pachet 1-D, in which also details are decomposed at a deeper level. In this experimental situation a decomposition at level 5 was necessary. From the complete tree, the best-tree with a Shannon entropy criterion was computed. At this step the denoising of the signal was calculated using a global similar soft amplitude threshold.

Results and Discussion

Whole-cell inward Cl⁻ currents were recorded at membrane potentials between 0 and -120 mV in cultured cortical astrocytes under control conditions. Outside-out recordings on the same cells did not allow to resolve clear single channel events. In Figure 1A, single-channel current traces of 3500 sampling points, part of longer records, obtained in response to a test potential of -100 mV from an holding potential of 0 mV are shown either in control condition or in the presence of zinc. The signals were basically due to white noise and no particular differences were observable comparing the control current traces and those after zinc application. Conversely, power spectra clearly depicted, in the low frequency domain (up to 20 Hz), a significant difference between the current traces with a decrease of the spectral density in the presence of zinc. In Figure 1B, power spectra of the records of Fig. 1 are illustrated. The traces used for the power spectra were filtered by means of Wavelet functions. We decided to use the Haar decomposition, in order to reconstruct the single-channel current activity. The results indicate that in the control traces, negative abrupt transitions in the low frequency domain similar to single-channel events could be clearly detected and that these channel transitions were modified in the presence of external zinc.

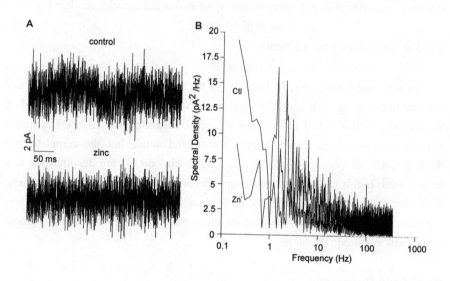

Figure 1. A) Currents traces recorded in outside-out patch-clamp configuration at a membrane potential of -100 mV in control conditions and after 2 min application of external zinc (1 mM). B) Power spectra of the same records show, in the low frequency domain, significative difference of the spectral density. Sampling time = 30 μs, number of points = 3500

The channel activity drastically decreased after application of zinc and a complete recovery was observed after wash-out. A representative signal decomposition in control situation and after zinc application (8000 points) is shown in Fig. 2. In control traces the presence of two channel levels could be resolved whereas no channels openings were observable in the presence of zinc. Similar analyses were performed at different membrane potentials (data not shown). In particular, at 0 mV no channel transitions were observed.

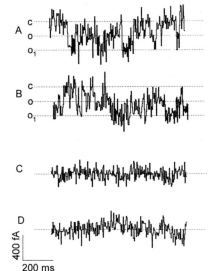

Figure 2. Current traces resulting from Wavelet transform decomposition. A,B) In control conditions the gating transitions of two channels have been detected. Closed and open states are indicated. C,D) In the presence of zinc no channels were active. Number of points = 8000

Taken together these results show that a small-conductance Cl^- channel, unresolvable from noise in conventional single-channel patch-clamp experiments, with a unitary Cl^- chord conductance of about 2 pS, underlies to the macroscopic high inwardly rectifying Cl^- conductance.

References

Ferroni, S., C. Marchini, P. Schubert and C. Rapisarda (1995) *"Two distinct inwardly rectifying conductances are expressed in long-term dibutyryl-cyclic-AMP treated rat cultured cortical astrocytes"*, FEBS Lett. 367:319-325.

Ferroni, S., C. Marchini, M. Nobile and C. Rapisarda (1997) *"Characterisation of an inwardly rectifying chloride conductance expressed by cultured rat cortical astrocytes"*, Glia in press.

Pusch, M., K. Steinmeyer and T. Jentsch (1994) *"Low single channel conductance of the major skeletal muscle chloride channel, CIC-1"*, Biophys. J. 66:149-152.

Woll, K.H., M. D. Leibowitz, B. Neumcke and B. Hille (1987) *"A high-conductance anion channel in adult amphibian skeletal muscle"*, Pflugers Arch. 410:632-640.

BROWNIAN MOTION AND THE ABILITY TO DETECT WEAK AUDITORY SIGNALS

ILSE CHRISTINE GEBESHUBER
*TU-BioMed, University of Technology, Wiedner Hauptstr. 8-10/1145,
1040 Wien, Austria*
E-mail: igebes@fbma.tuwien.ac.at
and
A. MLADENKA, F. RATTAY, W.A. SVRCEK-SEILER
TU-BioMed, Wiedner Hauptstr. 8-10/1145, 1040 Wien, Austria

ABSTRACT

The auditory system is of remarkable sensitivity: the weakest signals which can be detected cause in the inner ear vibrations of only some 100 pm. A model of the Brownian motion of the hairs of inner hair cells, which are responsible for the transduction of the mechanical sound signal into action potentials, is combined with an electrical model of the inner hair cell. It yields the spiking pattern of auditory nerve fibers. The spiking times are the input for an artificial neural network which determines whether there is a signal. The simulation shows that the spontaneous spiking activity of an auditory nerve fiber, which is partly caused by Brownian motion, enhances the possibility to detect weak auditory signals. For example, sampling enough information to recognize a tone with a signal to noise ratio as low as 1/10 in the spiking pattern of 100 auditory nerve fibers connecting to adjacent hair cells takes about 20 ms. Without the contribution of Brownian motion, the weak signal would not be detectable.

1. Introduction

Acoustical signals cause pressure waves of the inner ear fluids and thereby the motion of biological structures. About 3500 inner hair cells (IHC) are mounted in a line on the top of the basilar membrane, a 35 mm long elastic structure, and detect every motion within the hearing organ. When a pure sinusoidal tone is presented to the ear, only the hairs (stereocilia) of a rather small portion of the IHC vibrate significantly, which is a consequence of the mechanical properties of the basilar membrane which changes along its length.

The hair cell transforms the mechanical signal into a neural one (mechanoelectric transduction). The movement of the stereocilia controls the ion current flow into the cell, i.e. the time course of the intracellular potential of the hair cell reflects the motion of the hairs. Intracellular voltage changes in the range of only 0.1 mV are able to cause neurotransmitter release at the synaptic contacts at the bottom of the hair cell and neural impulses in the fibers of the auditory nerve (Hudspeth, 1989).

The ear is a very sensitive organ: it is able to detect signals which lead to vibrations of the stereocilia with amplitudes of only 0.3 nm (Hudspeth, 1989). The stereociliary movement caused by Brownian motion is about 3.5 nm RMS (Denk et al., 1989) and is in part responsible for the spontaneous activity: without any acoustic stimulus, firing rates up to 160 spikes/s have been measured in auditory nerve fibers (Relkin and Doucet, 1991). The first neural responses at barely threshold intensities appear to be a decrease in spontaneous activity, and phase locking of spike discharges to the stimulus cycle. This is not to say that the fiber will fire in response to every cycle of a near-threshold stimulus. Rather, even though the overall discharge rate may not be significantly greater than the spontaneous level, those spikes that do occur will tend to be locked in phase with the stimulus cycle (Hind, 1972).

The aim of this study is the computational demonstration that Brownian motion of inner hair cells stereocilia enhances the ability to detect low level auditory tones from auditory nerve spiking patterns.

2. Materials and Methods

The stereocilia of one IHC are interconnected by links (Pickles et al., 1984). Therefore, the hair bundle behaves like a system of coupled overdamped harmonic oscillators. The bundle motion is simulated by numerical integration of a system of coupled Langevin equations, i.e. Newtonian equations of motion with stochastic inhomogeneities (Chandrasekar, 1943). The parameters used in the simulation are stereocilia stiffness, mass and damping constants (Steele and Jen, 1988) and link stiffness and pre-tension (Zetes, 1995). The frequency distribution of the displacements caused by Brownian motion is of Lorentzian shape with a knee frequency of about 400 Hz. Close inspection of the spectral density shows high frequency components due to the bundle's inner degrees of freedom (Svrcek-Seiler et al., 1997).

Both the deterministic movement induced by the stimulating signal and the stochastic contribution of Brownian motion cause the total displacement of the stereocilia. The equivalent electric circuit model of an IHC has three different compartments (stereocilia, cell body, transition part), each compartment consisting of capacitance, resistance and battery to model the various currents through the membrane, which are caused by the typical ionic concentrations of the fluids surrounding the compartments (physiological parameters used for modeling taken from Zenner, 1994, pp. 81-230). The current flux through the 60 transduction channels per IHC is modeled individually by subcompartments containing switches, which simulate the displacement sensitive opening/closing kinetics of the transduction channels. The compartments are connected by resistances, which represent the electrical properties of the cytoplasm. The currents from the open transduction channels combine in a current which flows into the IHC, leads to receptor potential changes in the cell body and leaves the receptor cell through ion channels in the basolateral cell body membrane.

A threshold function computes the spiking times in the nerve fibers: A receptor potential change of 0.1 mV is assumed to induce the first spike and a jump of the threshold voltage to a high value, followed by exponential decrease to its resting state at 0.1 mV above the resting potential. When the threshold voltage function again crosses the IHC voltage, the next spike arises (Fig. 1). Even a highly sensitive auditory nerve fiber with a high spontaneous rate can only fire with a maximum rate of some 100 spikes/s and thence, many of the IHC voltage peaks are not represented. Because an average of 8 fibers connect to each IHC, voltage peaks which are not found in a specific fiber may be found in one of the others.

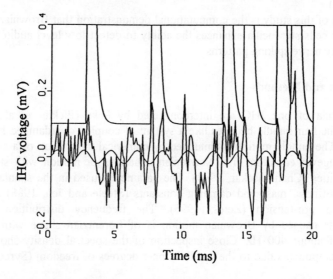

Figure 1: The stimulating 500 Hz tone (sinusoidal curve) and the resulting voltage changes in the IHC (zig- zag curve). The exponentially shaped "threshold curve" yields the firing times in a single auditory nerve fiber. The voltage changes theoretically evoked by the sinusoidal signal alone are too small to cause spiking and only with the help of noise the weak signal can be represented in the neural pattern.

An important temporal information of a single fiber is the sequence of its interspike times. The distances between the peaks in the interspike time histograms (ISTH) are closely related to the period duration of their stimuli. With increasing signal to noise ratio (SNR), the ISTH becomes gradually more regular, i.e. finally the time between the peaks of the histogram becomes multiples of period duration (Fig. 2).

We use MATLAB backpropagation neural nets with 16 input neurons, 1 output neuron and 2 hidden neurons in order to determine whether there is a signal. If the values from the ISTH are used as input data, the recognition rate is rather limited.

Therefore, a small set of simple input data is prepared: the first peak of the tested 500 Hz signal is expected at 2 ms, the first valley at 3 ms, the second peak at 4 ms etc. The bins corresponding to the time interval [1.5 ms - 2.5 ms] are added to value 1, the sum of the next bins corresponding to the time interval [2.5 ms - 3.5 ms] results in value 2, ... In this way, we can expect that even a disturbed 500 Hz signal will tend to generate a sequence of values going up and down, with high values for the index 1,3,5,7, ... and low values for the index 2,4,6,8, ... These data are used to produce a new sequence of 0 and 1: if an element of the old sequence follows a smaller one, a 1 is generated, else a 0. In this way, a strong sinusoidal tone will produce the sequence 1,0,1,0,1,0,1,0,1,0,1,0,1,0 - but a very weak signal will cause an essentially different code.

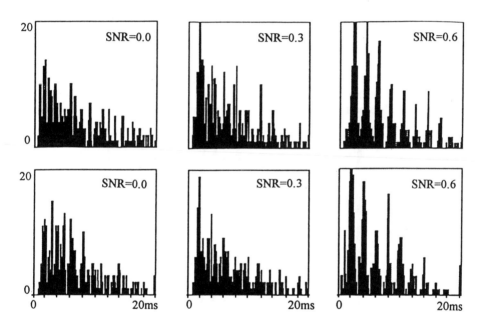

Figure 2: ISTH of a high spontaneous rate fiber. The SNR is varied for a 500 Hz sinusoidal stimulus from 0 to 0.6. The stimulus is presented for 2 s and generates about 200 spikes for every run. Binwidth 0.2 ms, only interspike times up to 20 ms are shown. The upper and lower traces show the results from two runs, the differences in the ISTH are caused by the stochastic portion of the stimulating signal.

The success of this method depends essentially on signal duration. This means that it is more difficult to discern auditory signals when they are presented only for a short time, because the information is less regular for short signals. This fact is in accordance with experimental results (e.g. Mark and Rattay, 1990) and is also seen in Fig. 3: long signals hold the 1,0,1,0,... structure even for weaker tones.

3. Results

Figure 1 shows the stimulating 500 Hz sinusoid and the resulting voltage changes in the IHC receptor potential. As the contributions of the noise caused by the Brownian motion of the stereocilia are one order of magnitude above the ones of the weak signal, the receptor potential changes are very noisy. The exponentially shaped threshold curve simulates the resulting spiking pattern in an auditory nerve fiber.

From the spiking pattern, the ISTH are calculated. Figure 2 shows ISTH for a weak 500 Hz tone with SNR from 0 to 0.6. At an SNR of 0.6, the intervals between spikes are very close to multiples of the 2 ms period duration. The 2 ms periodicity in the peaks of the ISTH is more difficult to decipher for SNR=0.3 (middle) and, of course, it is lost in the left traces where there is no periodicity in stimulus. Top and bottom traces show the results of two runs, the differences are caused by the stochastic nature of the stimulus.

Figure 3 shows the 1-0 strategy input vectors for the neural network which result from treating the ISTH with the method described above. The sequences of zeros and ones become more and more regular with increasing stimulus presentation time. This holds even for very weak sinusoidal signals with a SNR of 0.1.

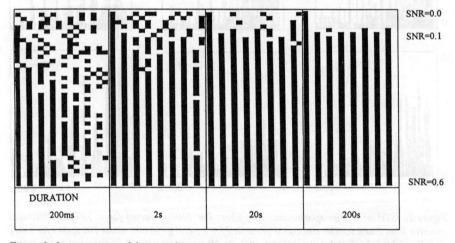

Figure 3: Input patterns of the neural network using the 1-0 strategy for different durations of the acoustic signal. Each pattern consists of 41 lines with 16 numbers each. Black squares mean input equal to 1, white squares mean input equal to 0. The SNR increases from the top to the bottom. The simulated receptor potential deviations from the resting potential caused by the deterministic sinusoidal signal alone vary from 0 to 0.02 mV in steps of 5.10^{-5} mV. The regularity of the simplified neurogram increases with the duration of the acoustic signal. However, even for very long durations (200 s) auditory signals with an SNR essentially smaller than 0.1 cannot be detected, because the periodicity information is lost. Note that the irregularities caused by noise, i.e. for SNR=0, can be seen for 200 ms to 20 s duration, but not in the 200 s signals.

The MATLAB backpropagation network is trained with input patterns belonging to acoustic stimuli of 500 Hz which cause receptor potential changes from 0 to 0.08 mV (when neglecting the influence of Brownian motion) and different stimulus presentation times ranging from 200 ms to 200 s. For every signal duration 100% recognition is obtained for deterministic receptor potential changes greater than 0.02 mV. At a stimulus presentation time of 20 s, the artificial neural network always recognizes even the 5.10^{-5} mV signal, the smallest nonzero case tested. But even for shorter signal presentation times, the recognition rates for signals with a SNR>0.1 are 90% and more.

4. Discussion

The acoustical information enters the biological neural net in form of spikes, essentially disturbed by the spontaneous activity which is in part caused by the Brownian motion. A single fiber needs about 2 seconds to sample enough information to decipher a tone with a signal to noise ratio of 1/10. This result is equivalent to 20 ms signal presentation time, if one considers 100 fibers connecting to adjacent IHC, since the Brownian motion of harmonic oscillators is an ergodic process and the noise sources to independent fibers are assumed to be independent.

This fact seems to indicate that the threshold of hearing is reached when the amplitude of the stereociliary movement is one order of magnitude below that caused by Brownian motion. Given the threshold of 0.1 mV receptor potential change for neurotransmitter release, the intensity of the voltage fluctuations induced by the Brownian motion of the stereocilia are in a range that allows to make signals with an intensity one order of magnitude lower detectable. If the stochastic voltage changes were considerably smaller or larger, the human hearing threshold for weak tones would be worse. This result can therefore be related to the phenomenon of noise-induced linearisation (Spekreijse and Oosting, 1970) or stochastic resonance (Benzi et al., 1981), for which a simple model is a symmetric bistable or threshold process x(t) driven by both an additive random noise, and an external sinusoidal bias. On keeping the forcing amplitude and frequency fixed, the amplitude of the periodic component of the process, x, grows sharply with the noise intensity until it reaches a maximum and, then, decreases slowly according to a certain power-law.

In this multicellular model for peripheral auditory coding we have demonstrated the enhancing effect of Brownian noise on the high sensitivity of the auditory system and have shown that the noise quantitatively accounts for the detection of weak pure tones, using a simplified description of the transduction channel kinetics, neurotransmitter release mechanisms and the mechanical properties of the coupled stereocilia. We believe that these effects are robust and will also hold for more biophysically detailed models of the transduction process; the extension of this to complex and broad band auditory signals, such a clicks and speech, is an open question.

References

Benzi R., A. Sutera and A. Vulpiani (1981), The mechanism of Stochastic Resonance, *J. Phys. A* 14L: 453-457.

Chandrasekar S. (1943), Stochastic problems in physics and astronomy, *Rev. Mod. Phys.* 15: 21-44.

Denk W., W.W. Webb and A.J. Hudspeth (1989), Mechanical properties of sensory hair bundles are reflected in their Brownian motion measured with a laser interferometer, *Proc. Nat. Acad. Sci.* 86: 5371-5375.

Hind J.E. (1972), Physiological correlates of auditory stimulus periodicity, *Audiology* 11: 42-57.

Hudspeth A.J. (1989), How the ear's works work, *Nature* 341, 397-404.

Mark H.E. and F. Rattay (1990), Frequency discrimination of single-, double- and triple-cycle sinusoidal acoustic signals, *J. Acoust. Soc. Am.* 88: 560-563.

Pickles J.O., S.D. Comis and M.P. Osborne (1984), Cross links between stereocilia in the guinea pig organ of Corti, and their possible relation to sensory transduction, *Hear. Res.* 15: 103-112.

Relkin E.M. and J.R. Doucet (1991), Recovery from prior stimulation. I: Relationship to spontaneous firing rates of primary auditory neurons, *Hear. Res.* 55: 215-222.

Spekreijse H. and H. Oosting (1970), Linearizing: a method for analysing and synthesising nonlinear systems, *Kybernetik* 7: 22-31.

Steele C.R. and D.H. Jen (1988), Mechanical analysis of hair cell microstructure and motility, *Cochlear Mechanisms, Structure, Function and Models* (Ed. J.P. Wilson and D.T. Kemp), Plenum Press, New York.

Svrcek-Seiler W.A., T. Biró, I.C. Gebeshuber, F. Rattay and H. Markum (1997), Influence of Brownian motion on the detection of low-level acoustic tones, submitted to *J. Theor. Biol.*

Zenner H.P. (1994), Physiologische und biochemische Grundlagen des normalen und gestörten Gehörs, *Oto-Rhino-Laryngologie in Klinik und Praxis* (Ed: J. Helms), Georg Thieme Verlag, Stuttgart, Vol. 1: Ohr.

Zetes D.E. (1995), Mechanical and morphological study of the stereocilia bundle in the mammalian auditory system, PhD Thesis, Stanford University.

DELAYED LUMINESCENCE IN BIOPHYSICAL INVESTIGATIONS

A. SCORDINO, A. TRIGLIA, F. MUSUMECI
Institute of Physics and INFM Unit, University of Catania
Viale A. Doria 6, I-95125 Catania (Italy)
E-mail: demone@if.ing.unict.it

and

R. VAN WIJK
Department of Molecular Cell Biology, University of Utrecht
Padualaan 8, 3584 CH Utrecht (The Netherlands)
E-mail: r.vanwijk@biol.ruu.nl

ABSTRACT

Decays of Delayed Luminescence emitted in the seconds range from the giant unicellular alga *Acetabularia acetabulum* are measured. Results show a correlation between changes in delayed luminescence and chloroplast motility, emphasizing the possibility of using delayed luminescence measurements as an intrinsic probe in biophysical investigations.

1. Introduction

The phenomenon of light emission from green plants seconds after illumination was observed from Arnold and Strehler in 1950. This postillumination emission of light has been referred in literature by a number of terms: delayed fluorescence, delayed light emission, delayed luminescence and luminescence (Jursinic, 1986). We know now that the delayed light is something like $10^3 \div 10^5$ times lower in intensity than is the fluorescence, and it is typical not only of green plants (Veselova *et al.*, 1988; Chwirot & Popp, 1991; Van Wijk *et al*, 1993). In the following we will use the term Delayed Luminescence, shortly indicated as DL.

DL is by nature extremely polyphasic; the lifetime spectrum extends from 10^{-7} to more than 10 s (Lavorel *et al.*, 1986). So from measurements of DL one may gather information about energy transfer pathways from more distant molecules, taking into account long-range interactions.

Some phenomenological results that we obtained using different species (Scordino *et al.*, 1996; Musumeci *et al.*, 1994) let to think that DL (that we measure from 100 ms up to 100 s) could become an important intrinsic probe for measuring the physiological state of the sample. It can be supposed that these intriguing luminescence data reflect the general functional order that the biological organisms show in addition of their very complex chemical and structural composition. Understanding this order is one of the challenging problem today.

In this respect it is important to study living structures as a whole because, for instance the occurrence of collective phenomena are absent after extraction of subsystems.

From this point of view an interesting species for research is the giant unicellular green alga *Acetabularia acetabulum*. This organism, whose dimension may vary considerably (up to a few centimeters long), has the advantage of being a single cell organism that can be easily grown in the laboratory, so that it is possible to obtain several samples with uniform characteristics.

We carried out measurements on *Acetabularia* samples in order to provide comparative experimental information about possible correlation between DL characteristics and biological activity which reflects a major characteristic for functional order, as for instance organelle motility.

Organelle motility occurs in a strictly organized manner. In fact it has recently been discovered (Menzel, 1994) that the majority of chloroplasts in *Acetabularia* are interconnected by long tubular membrane bridges. There are indications that the tubules not only connect the chloroplasts but actively take part in the motile mechanism, interact with the cytoskeleton and effectively modify chloroplast behavior.

2. Materials and Methods

Measurement consists of illuminating the biological sample and of counting the number of photons re-emitted from the sample after the light source had been switched off.

The experimental set up (Scordino *et al.*, 1996) consists of a dark chamber where the samples are maintained at a constant temperature. The emitted radiation is detected by a low-noise photomultiplier cooled down to -20 °C, working in single photon counting mode. The illumination of the samples is performed using a flash lamp (METZ 45CL1) in order to have short duration (about 3 ms). Due to the time lag of the experimental set-up, the photon counting starts some tens of ms after the source is switched-off.

Every sample for measurement is constituted of a single or a few algae put in a plastic Petri dish (6 cm in diameter) containing artificial seawater. For each sample a series of three or five runs has been performed, whose average value was considered. The background emission, due to the DL of the empty cuvette, has been measured in the same experimental conditions and subtracted from the data obtained from each sample. Random noise, which becomes dominant at long times of the DL curves, was reduced by processing raw data with a smoothing procedure (Scordino *et al.*, 1996) such that the sampled data were equally spaced on a logarithmic time scale.

Acetabularia acetabulum (L.) samples were kindly given by Dr. G.Thiel of Pflanzenphysiologisches Institut der Universität Göttingen.

3. Results and Discussion

For all the samples examined, the experimental data of the decay of delayed luminescence in the time range of our measurements fit to the hyperbolic trend as:

$$I(t) = \frac{A}{(t+t_0)^m} \tag{1}$$

The parameters A, t_0 and m which characterize Eq.1 have been calculated adapting to the specific problem the general mathematical method of nonlinear least-square fitting. Reduced χ^2 figures less than 2 have been obtained. The reliability of the parameter estimation was tested comparing several measurements repeated on the same sample: it was found that the ratio between the standard deviation and the mean value obtained in the different measurements was less than 5%.

The hyperbolic form Eq.(1) has already been used in the literature to describe the decay of delayed luminescence. Different hypotheses have been formulated in order to explain such a trend as the existence of biequimolecular processes (Jursinic, 1986), of excited levels whose decay rates are distributed according to a suitable probability density function (Scordino et al., 1996), of a coherent field, present inside the biological matter, which plays a relevant role in the maintenance of cell organization (Popp et al., 1988; Popp & Li, 1993).

Preliminary results have shown that the slope of DL decays from *Acetabularia acetabulum* samples strongly depends on the intensity of illumination. More precisely, on decreasing the excitation the parameter m of Eq.1 decreases, while over a certain threshold value of the intensity characteristic oscillations appear in the decay trend. So it seems that the intensity of illumination influences not only the number of excited levels but also the mechanism of relaxation. This non linear behaviour suggests that non local effects are present and that DL could be deeply connected with the basic structure of the cell and, in particular, with its cytoskeleton.

To test if this is true experiments in two different conditions which affect organelle motility have been performed, as explained in the following.

3.1. Effects of immersion in liquid nitrogen

A sample, constituted by three algae of the same length, was immersed in liquid nitrogen for 15 sec and then tested in normal artificial seawater at 20 °C. Measurements of DL were performed at initial standard condition, after immersion in liquid nitrogen and successive immersion in normal seawater at 20°C, 1 hour later and two days later. The corresponding decays are reported in Fig.1. It appears that after the immersion the trend is about parallel to the initial unperturbed trend, but considerably lower even if the sample appeared "green" colored. After a few days, in any case, the alga became completely white colored, while the residual signal disappeared.

Observations at microscope on samples which suffered the same stress conditions showed that the algae remained intact, but no evidence of streaming of chloroplasts, as observed in standard condition, could be revealed. The organelles appeared to be frozen and only some of them, which have more room nearby, showed a vibrational motion round a fixed position. No indication of any recovery within one hour was observed.

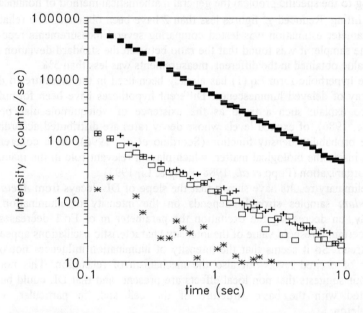

Figure 1 - Delayed Luminescence from *Acetabularia acetabulum* samples: liquid nitrogen immersion test. (■) Start (20°C, seawater), (+) after immersion in liquid nitrogen, (□) 1 hour later, (∗) 48 hours later. All measurements were performed in seawater at 20°C.

3.2. Effects of chloroform

A sample constituted by one alga was immersed in seawater with chloroform (final concentration of about 1 mM of chloroform) for 15 minutes and then tested in normal chloroform-free conditions. Adding of chloroform was performed after cooling at 3°C in order to reduce chloroform volatility. Measurements of DL were performed at initial standard condition, after reaching thermal equilibrium at 3°C, 10 minutes after the addition of chloroform solution, immediately after restoring standard conditions (chloroform-free seawater at 20°C) and 4 hours later. Results are reported in Fig.2 where each curve is the average of the trends of four different samples examined at the same time. Figure 2 shows a progressive restoring of the DL decays.

Observations at microscope on samples which suffered the same stress conditions showed that at 3°C streaming was slightly slower than in standard condition. After the addition of chloroform organelles moved progressively slower and in about five minutes the streaming was frozen. If at this point the algae were put in fresh seawater at 20°C, in about 3 minutes the streaming was recovered. Normally the algae appeared in good conditions at the end and survived.

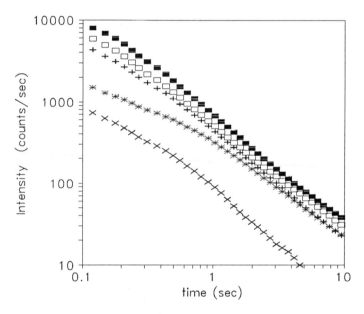

Figure 2 - Delayed Luminescence from *Acetabularia acetabulum* samples: chloroform solution immersion test. (■) Start (20°C, seawater), (∗) after cooling (3°C, seawater), (×) 10 min after addition of chloroform (3°C, chloroform solution), (+) immediately after restoring standard conditions (20°C, fresh seawater), (□) 4 hours after restoring standard conditions.

4. Conclusions

Results obtained emphasize the fact that the study of DL of a whole organism could provide information that could not be obtained from its subsystems after isolation (as for instance isolated chloroplasts in this case).

Results obtained with chloroform in particular show that DL measurements may provide information on the role of cytoskeleton, and of the microtubules in particular, on organized behavior of cell organelles.

The role of microtubules, in particular is in debate as it regards the action of anesthetic substances. In fact (Penrose, 1995, pp.367-371), due to the fact that general anesthetic substances seem to have no chemical relationship with one another, interactions different from simple enzyme inhibitions have been hypothesized as responsible for anesthesia. It has been suggested that general anesthetics may act through the agency of their van der Waals interactions with the conformational dynamics of the tubulin (dimers) proteins in microtubules.

The role of microtubules is also discussed as it regards the Fröhlich's ideas for the possibility of quantum-coherent phenomena in biological systems. Fröhlich's original ideas seem to have been that such large-scale quantum states would be likely to occur in cell membranes, but it has been suggested to seek such quantum behavior in the microtubules.

The correlation shown by the above results suggests the possibility of using DL measurements in order to closely investigate the biophysics of a living organism as a whole complex system.

References

Chwirot B., F.A. Popp (1991): "White light induced luminescence and mitotic activity of yeast cells", *Folia Histochemica et Cytobiologica* **29**, 155.

Jursinic P.A. (1986): "Delayed fluorescence: current concepts and status", in *Light Emission by Plant and Bacteria*, J. Govindjee et al. eds , Academic Press, New York, 291-328

Lavorel J., J. Breton, M. Lutz (1986): "Methodological principles of measurement of light emitted by photosynthetic systems", in *Light Emission by Plant and Bacteria*, J. Govindjee et al. eds , Academic Press, New York , 57-98.

Menzel D. (1994): "An interconnected plastidom in Acetabularia: implications for the mechanism of chloroplast motility", *Protoplasma* **179** , 166-171

Musumeci F., A. Triglia, F. Grasso, A. Scordino, D. Sitko (1994): "Relation between delayed luminescence and functional state in soya seeds", *Il Nuovo Cimento D* **16**, 65-73

Penrose R. (1995): *Shadows of the Mind. A Search for Missing Science of Consciouness*, Vintage Science

Popp F.A., K.H. Li, W.P. Mei, M. Galle, R. Neurohr (1988): "Physical Aspects of Biophotons", *Experientia* **44**, 576-585

Popp F.A. and K.H. Li (1993): "Hyperbolic Relaxation as a Sufficient Condition of a Fully Coherent Ergodic Field", *International Journal of Theoretical Physics* **32**, 1573-1583.

Scordino A., A. Triglia, F. Musumeci, F. Grasso, Z. Rajfur (1996): "Influence of the presence of atrazine in water on in-vivo delayed luminescence of Acetabularia Acetabulum", *Journal of Photochemistry and Photobiology B: Biology* **32**, 11-17.

Van Wijk R., H. Van Aken, W.P. Mei, F.A. Popp (1993): "Light-induced photon emission by mamalian cells", *Journal of Photochemistry and Photobiology B: Biology* **18**, 75-79.

Veselova T.V., V.A. Veselovsky, V.I. Kozar and A.B. Rubin (1988): "Delayed luminescence of soybean deeds during swelling and accelerated aging", *Seed Science & Technology* **16**, 105-113.

PAINLEVE' PROPERTY OF ODEs AND TIME SERIES ANALYSIS

A. DI GARBO, M. BARBI, S. CHILLEMI
Istituto di Biofisica CNR, Via S. Lorenzo 26, 56127 Pisa (Italia)
E-mail: digarbo@ib.pi.cnr.it

ABSTRACT

The analytical properties of the solutions of a system of ODEs in the complex time plane influence its evolution on the real time axis. In particular, the extremes of the real solution can be associated with the singularities of the complex solution falling near the real axis. This result is used here as starting point to introduce an algorithm for testing time series nonlinearity.

1. Introduction

In the last years several efficient methods for the analysis of time series (Abarbanel *et al.*, 1993) have been proposed. The algorithm described in this paper is able to detect nonlinearity in a time series based on the analysis of the distribution of its extremes and the comparison with surrogate data. The main theoretical argument that underlies this algorithm is taken from the theoretical and numerical results obtained in the study of systems of ordinary differential equations (ODEs). It has been shown that the dynamical behaviour of an ODE's solution is strongly connected to its analytic properties in the complex time plane and in particular to the singularities near the real axis (Ramani *et al.*, 1989). Those results suggest that this algorithm should be able to distinguish between the distributions of extremes that can be accounted for by linear stochastic processes and those that can not (Theiler *et al.*, 1992).

2. The algorithm

As recalled above, for solutions of ODEs, a strong connection between the dynamical properties of the real time evolution and the analytical ones in the complex time plane has been shown to exist (Ramani *et al.*, 1989). For instance, the solutions of the Lorenz equations in the complex time plane were studied analytically and numerically for different dynamical regimes (Tabor & Weiss, 1981). It was found that the distribution of the singularities in the complex time plane determines completely the corresponding real time behaviour of the solution. The main contributions come from the singularities nearest to the real time axis. For example in the periodic regime of the Lorenz equations (limit cycle) the distribution of the singularities (poles) is regular and this reflects the corresponding periodicity of the real solution. As the system approaches the chaotic regime, the corresponding distribution of singularities becomes very irregular. Moreover it was shown numerically that the positions of the singularities nearest to the real time axis are associated with the local maxima and minima of the real time solution. Similar results were obtained also for the Duffing equation (Konno & Tateno, 1984). Moreover, the distance of the nearest singularities to the real axis was shown to be connected to the real values of the solution (Konno & Tateno, 1984; Tabor & Weiss, 1981). The conclusion that arises naturally is that the pattern of the extremes of the real time solution contains dynamical information. A further point to be stressed here is that for a system of ODEs the position of the singularities (if they exist) is fixed deterministically (Ramani et al., 1989).

With these results in minds, the recipe for the algorithm is the following:
a) consider the set of couples $\{t_j, y(t_j): j=1, n\}$ where $y(t_j)$ are the local maxima (one could likewise use the local minima);
b) compute the quantity

$$L = \sqrt{\sum_{j=1}^{n-1} [(t_{j+1} - t_j)^2 + (y(t_{j+1}) - y(t_j))]^2} \qquad (2.1)$$

that represents the length of the broken line joining the points $\{t_j, y(t_j): j=1, n\}$;
c) test whether the value of L can be statistically accounted for by assuming the time series to be generated from a linear stochastic process. To this end ten Fourier-shuffled surrogate series consistent with this null hypothesis are created and the L statistic is computed on them too.

More details on the surrogate data properties can be found elsewhere (Theiler et al., 1992). In the next section we will show the results obtained by applying this new algorithm to several artificial time series.

3. Results and conclusions

Time series were generated from well known models in order to test the algorithm and find its possible limitations. The Lorenz equations

$$\begin{aligned} \dot{x} &= \sigma(y - x) \\ \dot{y} &= \rho x - y - xy + \varepsilon \xi(t) \\ \dot{z} &= -\beta z + xy \end{aligned} \qquad (3.1)$$

where parameters are set to the standard values $\sigma = 10$, $\rho = 28$, $\beta = 8/3$ for the chaotic regime and $\xi(t)$ is gaussian white noise of unity standard deviation, were integrated numerically; the integration step was $h = 0.001$ and the sampling time was $\Delta t = 0.025$. By sampling $z(t)$, two time series were obtained, respectively for the values $\varepsilon = 0$ and $\varepsilon = 0.2$. Two time series were generated from the Henon map, with and without additive noise. Two more time series were obtained from the deterministic signal $x(t) = sin\,(0.1\ t) + \varepsilon\,\xi(t)$, for $\varepsilon = 0$ and $\varepsilon = 0.3$ respectively; the sampling time for both cases was $\Delta t = 0.15$. Two final time series represented Gaussian white noise and coloured noise.

To determine the local maxima (or minima) in the time series $y(t)$ the first difference time series $y(t + 1) - y(t)$ is simply inspected for changes of sign. Once the length L defined by Eq. (2.1) is calculated on the original series and on the surrogate one, the difference of the two values is statistically checked.

The results obtained by using this algorithm are shown in Fig. 1, plotting the values of the significance parameter $S = |L - L_s|/\sigma_s$ for the eight time series. For points above the dashed line ($S = 2$) the null hypothesis of a linear stochastic process can be rejected (Theiler *et al.*, 1992) with a probability larger than 0.95. The values of L were calculated by using the local maxima.

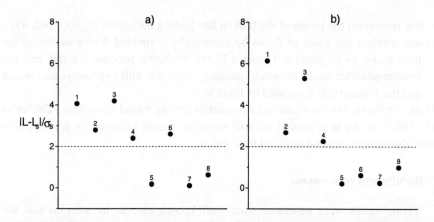

Fig. 1 Significance of nonlinearity measured for the model time series: L is calculated on the original data while L_s is the mean value obtained on the surrogate data and σ_s is the corresponding standard deviation. The correspondence is the following: 1) Lorenz equations; 2) Lorenz equations plus noise; 3) Henon map; 4) Henon map plus additive noise $\varepsilon = 0.35$; 5) sinusoidal wave; 6) sinusoidal wave plus noise; 7) Gaussian white noise; 8) Coloured noise $x(t+1) = a\,x(t) + \varepsilon\,\xi(t)$, $a = 0.85$, $\varepsilon = 0.2$, $\xi(t)$ is Gaussian white noise. Number of points used: a) $N = 1000$ b) $N = 2000$.

Figure 2 shows the results (corresponding to the right panel of Fig. 1) obtained from Eq. (2.1) by using the local minima.

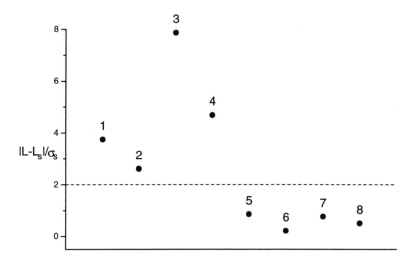

Fig. 2 As for Fig. 1 but using the local minima; $N = 2000$.

Let us comment on these results. We have seen in the previous figures that the time series corresponding to chaotic dynamics (time series : 1,2,3,4) exhibit the expected nonlinear character. Similarly time series #7 and #8, corresponding, respectively, to Gaussian white noise and coloured noise, present values clearly lower than 2. This means that the null hypothesis they are generated by linear stochastic processes cannot be rejected with probability higher than 0.95. Coming to the case of time series #6, corresponding to the sine wave plus noise, we see in Fig. 1a that it exhibits an unexpected nonlinear character. However, as the number of points is doubled (Fig. 1b), this anomaly disappears. From all these facts it follows that : a) for chaotic dynamics the corresponding nonlinear character is detected by the algorithm; b) for coloured noise or gaussian white noise the linear character ($S < 2$) is confirmed; c) in the cases where the power spectrum of the signal contains a strong (dominant) peak superimposed on a broad band the algorithm leads to contrasting results as the number of points used is varied. This last conclusion can be explained by the fact that using surrogate data for signals with long coherence time leads to results that must be interpreted very carefully (Theiler et al., 1992). The results presented here are preliminary and further work is needed to better understand the performance of this algorithm.

References

Abarbanel, H. D., I. R. Brown, J. J. Sidorowich, L. S. Tsimring (1993) "The analysis of observed chaotic data in physical systems", *Rev. of Mod. Physics* 65: 1331-1392.

Konno, k and H. Tateno (1984) "Duffing's equation in complex time and chaos", *Prog. Theor. Phys.* 72: 1047-1049.

Ramani, A, B. Grammaticos, T. Bountis (1989) "The Painleve' property and singularity analysis of integrable and non-integrable systems", *Phys. Rep.* 180: 161-245.

Tabor, M and J. Weiss (1981) "Analytic structure of the Lorenz system", *Phys. Rev.* A24: 2157-2167.

Theiler, J., S. Eubank, A. Longtin, B. Galdrikian, D. Farmer (1992) "Testing for nonlinearity in time series: the method of surrogate data", *Physica* D58: 77-94.

EVALUATION OF SELF-SIMILARITY AND REGULARITY PATTERNS IN HEART RATE VARIABILITY TIME SERIES

MARIA G. SIGNORINI

Dipartimento di Bioingegneria, Politecnico di Milano, p.za Leonardo da Vinci, 32, Milano, 20133, Italy
E-mail:signorini@biomed.polimi.it

ABSTRACT

The paper presents results in Heart Rate Variability (HRV) signal analysis by measuring long period correlation and self-similarity properties and a time series regularity index in the short period. We compare three methods for the Hurst self-similarity exponent H computation: calculation of α exponent of $1/f^{\alpha}$ spectrum ($H=(\alpha-1)/2$), Absolute values of the aggregated series and Higuchi parameter. Approximate Entropy (ApEn) instead quantifies for short data sets (<1000) the degree of the series regularity. We study the HRV signals in the 24h of Normal, Heart Transplanted and Myocardial Infarction (MI) subjects. Results show that the H parameter is significantly different in healthy versus Heart Transplanted subjects ($p<10^{-7}$). ApEn values separate ($p<0.05$) MI patients who had reduced values of the ventricular ejection fraction. This reduction is associated to a higher risk for sudden cardiac death. Results confirm the ability of ApEn index to classify HRV patterns. Computation of H can be a valid help for diagnosis of cardiovascular pathologies.

1. Introduction

The introduction of methods estimating some parameters which are a measure of the invariant characteristics of nonlinear system attractors has allowed many experimental works to identify the presence of nonlinear chaotic dynamics in biological system regulation and control (Parker & Chua, 1987; West, 1990).

A large number of these approaches is based on the reconstruction procedure whose reliability is guaranteed by the Takens theorem (Abarbanel et al., 1993). Many others do not presume any a priori hypothesis about the existence of a system attractor with complex chaotic characteristics. They only perform a measurement of some geometrical properties of the time series which can be an index of the presence of nonlinear dynamics (Mandelbrot, 1985).

Frequency domain analysis of the HRV signal may provide a quantitative and noninvasive measure if the activity of the Autonomic Nervous System (ANS) (Kamath & Fallen, 1993). This is obtained by a linear modeling approach that quantifies both the sympathetic and parasympathetic control activity and their balance through the measure of spectral low and high frequency components (LF and HF) (Pagani et al, 1986). Recent results on HRV signal analysis show that its dynamic behavior involves nonlinear components that also contribute in the signal generation and control (Turcott & Teich, 1996; Cerutti et al, 1996). The aim of this paper is to investigate the structure of the physiological rhythms of the cardiac system involving nonlinear dynamics. We want to verify how the presence of nonlinear deterministic phenomena could affect the HRV signal both in short and long temporal windows.

2. Methods

2.1 Self-similarity H parameter

A time series $x(t)$ that satisfies (after appropriate setting of the initial conditions) the relation $x(ht) \stackrel{d}{=} h^H x(t)$ is said to be *self-similar*. The $\stackrel{d}{=}$ implies that both terms of the relation have the same distribution function. H is a constant ($0 < H < 1$), h is the time scaling factor. If a signal $x(t)$ is self-similar its distribution remains unchanged even after a rescaling of a factor h^H (Osborne & Provenzale, 1989).

H characterizes the level of *self-similarity*, giving us information on the recurrence rate of similar patterns in time at different scales. $H \cong 1$ implies a strong positive correlation, i.e. an increase of $x(t)$ is likely followed by another increase. $H \cong 0$ means a strong negative correlation: in this case a decrease (and vice-versa) follow an increase of x(t). If $H = 0.5$, then the self-similar signal $x(t)$ is completely uncorrelated. The self-similarity parameter is even known as *Hurst exponent, long-memory parameter, long-term correlation parameter* or simply *index H* (Mandelbrot, 1985).

We will now introduce three methods for the estimation of H. The two more recent ones (the *absolute value* and the *Higuchi* estimator indexes) have been compared to the traditional method based on the spectral analysis *(periodogram)* (Taqqu et al., 1996; Flandrin, 1989).

2.2 Estimation of H: Absolute values of the aggregated series

The method allows the H value to be estimated for a given time series $x(i)$ ($1 \leq i \leq N$, N length of the series) by computing first $X(i) = x(i+1)-x(i)$. The obtained $X(i)$ is divided into N/m blocks of size m. From the average of each block we obtain the aggregated series

$$X^{(m)}(k) = \frac{1}{m} \sum_{i=(k-1)m+1}^{km} X(i) \qquad k = 1, 2, ..., N/m. \qquad (1)$$

The *absolute values of the aggregated series* are:

$$M^{(m)} = \frac{1}{N/m} \sum_{k=1}^{N/m} |X^{(m)}(k)|. \qquad (2)$$

After repeating the procedure for different m values (m is the dimension of the data subset), we can plot $M(m)$ vs. m in log-log scale. $M(m) \propto m^{\gamma}$, where γ is related to the self-similarity parameter by the equation $\gamma = H - 1$ (Beran, 1994; Taqqu et al., 1996). Points of the log(M) versus log(m) plot should describe a straight line whose slope is γ. Figure 1a shows the index calculated for a fractional Brownian motion (fBm) with $H=0.3$, $H = \gamma + 1$.

2.3 Estimation of H: Higuchi's method

The first step is always the construction of a new time series $X^k{}_m$ from the original one $X(i)$ ($1 \le i \le N$, N length of the series, k initial series value, m reconstruction step).

The Higuchi estimator $L(m)$ evaluates the mean length of the new curve as a function of the m value (Higuchi, 1990).

$$\langle L(m) \rangle = \frac{1}{k} \sum_{k=1}^{m} L_k(m). \qquad (3)$$

If $\langle L(m) \rangle \propto m^{-\gamma}$, the original time series is fractal and its self-similarity parameter is given by $H = 2-\gamma$. The log-log plot of $\langle L(m) \rangle$ has a slope γ (Higuchi, 1988). Figure 1b shows an example of this behavior for a fBm ($H=0.3$).

2.4 Estimation of H : the periodogram

The computation of the periodogram yields an estimate of the power spectral density of the process, which is given by

$$P_X(f_m) = \frac{1}{N \Delta t} |X(f_m)|^2 = \frac{1}{N \Delta t} \left| \Delta t \sum_{k=0}^{N-1} x(k) \, e^{-j2\pi f k \Delta t} \right|^2, \qquad (4)$$

where $x(k)$ is the original time series and $X(f_m)$ its Fourier transform. If $x(k)$ is self-similar, its spectrum shows an inverse power-law $P_X(f) = \dfrac{1}{f^{\alpha}}$. α is the fractal exponent, which can be estimated as the slope of the log-log plot of $P_X(f)$. For fBm the self-similarity parameter is related to the fractal exponent by the equation $H=(\alpha-1)/2$ (Mandelbrot & Van Ness, 1968; Flandrin, 1989). The spectrum in Figure 1c shows a $1/f^{\alpha}$ behavior for many decades of f, with $\alpha = 1.6$, thus providing $H=0.3$ which is the expected value. For experimental time series some cautions should be

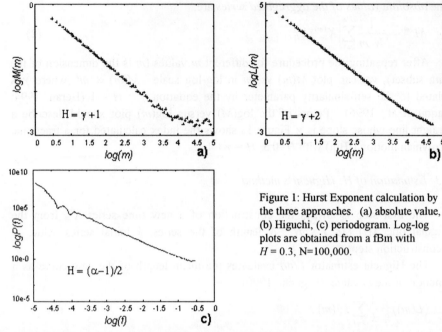

Figure 1: Hurst Exponent calculation by the three approaches. (a) absolute value, (b) Higuchi, (c) periodogram. Log-log plots are obtained from a fBm with $H = 0.3$, $N=100,000$.

applied: in fact the conversion from α to H is incorrect when $\alpha \leq 1$ and $\alpha \geq 3$, as the corresponding value of H is no more contained in the definition interval (West,1990).

2.5 Approximate Entropy

This is a statistic index which quantifies regularity and complexity and which appears to have potential application to a wide variety of relatively short (>100 values) and noisy time series data (Pincus, 1995). Its computation starts from the choice of two parameter m and r: m is the length of compared runs and r is really a filter. In practice we evaluate within a tolerance r the regularity, or frequency, of patterns similar to a given pattern of window length m.

From the numerical series $u(1)$, $u(2)$, ..., $u(N)$ and fixed $m \in N$,

ApEn(m, r)= avg. over i of ln[conditional probability that $|u(j+m)-u(i+m)| \leq r$ as $|u(j+k)-u(i+k)| \leq r$ for k=0,1, ..., m-1]. ApEn measures the likelihood that runs of patterns that are close for m observations remain close on next incremental comparison. It classifies both deterministic and stochastic signals requiring a reduced amount of points and is robust against noise contamination. The presence of regularity, i.e. the greater likelihood of the signal remaining close, produces smaller ApEn values and vice-versa. The value of N, the number of input data points is typically between 75 and 5000. Both theoretical analysis and clinical applications concluded that $m=1,2$ and r between 0.1 and 0.25 SD of the input data produce good statistical validity of ApEn(m, r, N). Figure 2 reports the ApEn values calculated in the 24 hours for a Normal subject. Parameters values are $m=1$, $r=0.2$, $N=300$.

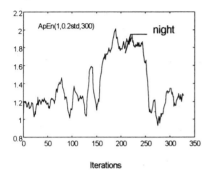

Figure 2: ApEn(1,0.2,300) calculated over 24h HRV signal of a normal subject. Differences in day and night epochs are shown.

3. Results

HRV signals come from 24 hours Holter recordings: we consider normal subjects (N), transplanted heart (T) and Myocardial Infarction patients.

In this section we try to find out if it is possible to distinguish healthy subjects from pathological ones only by studying the HRV series in terms of self-similarity parameter H or by the ApEn regularity index without making any assumption about the series generation process.

3.1 Normal subjects vs. patients with transplanted heart

We have analyzed with the three methods the 24 hours HRV series of 11 normal and 7 heart transplanted subjects. The length N of the RR interval series varies from 70.000 to 130.000 points.

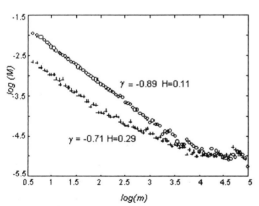

Figure 3 Absolute value of the aggregated series in a log-log plot for a normal subject (o) and for a heart transplanted patient (+).

Figure 3 shows an example of the absolute value M of the aggregated series for a healthy and for a heart transplanted subject. The healthy subject shows a higher value of the slope, which means a smaller value in terms of H. The regression line has been computed by excluding the first and the last decades of m (i.e. $1 \leq \log m \leq 4$). This choice removes noise disturbances in the regression interval estimation and it is adopted in all the analyses. In Figure 4, H values (avg.± std) of the two classes are reported as bar graphs for the three methods used together with the value of p provided by the

Student test. We obtained the ranges 0.12-0.22 with the *periodogram* ($p<0.02$), 0.17-0.31 with the *absolute value* ($p<2 \times 10^{-7}$), 0.13-0.21 with *Higuchi's* method ($p<0.0002$). *Higuchi's* estimator and the *absolute value* method exhibit the best performance, with very good values of p. Results are in agreement with other measurements obtained with different nonlinear parameters (Correlation Dimension, Entropy, and Lyapunov Exponents). They always confirm the role of HRV nonlinear dynamics as a global marker of cardiovascular pathologies (Guzzetti et al., 1996). Moreover preliminary results in ICU patients indicate that long period indexes seem to be strongly related to the prediction of both positive and negative patient outcome (Mainardi et al., 1997).

3.2 Myocardial Infarction subjects

The ApEn estimation performs a significant separation in the considered groups. Figure 5 reports ApEn values for 18 (9 + 9) MI patients with normal and reduced EF ($N=300$, $m=2$, $r=0,2std$ in *night* epoch). The analysis was repeated also for $N=1000$ and $N=3000$. The ApEn values always separate the two groups ($p<0.05$). ApEn is in the range 1.04 ± 0.2 - 0.95 ± 0.1 (reduced EF patients) and 1.31 ± 0.25-1.03 ± 0.07 (normal EF patients). During *day* epoch differences are even more evident with $p<0.01$. Data were evaluated during day and night period as already done for Correlation Dimension (D_2), H parameter and spectrum slope calculation (Lombardi et al., 1996). For this MI population, ApEn is able to separate normal and reduced EF patients with less computational effort.

4. Discussion

The results confirm the long memory characteristics of the HRV time series giving at the same time a strong indication for a possible clinical application of the calculated parameters. In particular by using ApEn it has been possible to classify MI patients who had different performances of the cardiac pump with a higher connected risk for sudden cardiac death.

The method of the absolute value is able to distinguish between transplanted and normal subjects with valuable results of the Student test. This is an important confirmation that a healthy cardiovascular system is strongly characterized by fractal properties in the time behavior of its variables. A cardiovascular system illness vice-versa produces a reduction of these complex patterns.

The results also suggest that these parameters could be helpful to integrate the diagnostic information on the heart pathology.

Complexity in the cardiovascular control can be related to a nonlinear model driving the system dynamics. The knowledge of these system properties introduces a new point of view in the heart pathophysiology together with more sensitive predictive parameters.

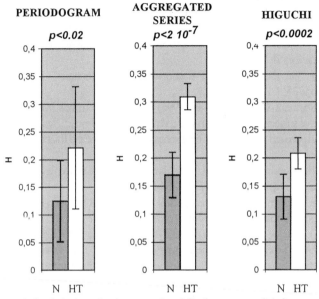

Figure 4: Statistical results (avg. ± std and Student test results) for normal and heart transplanted subjects.

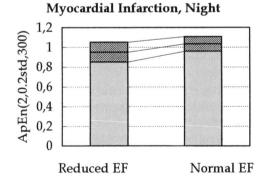

Figure 5: Approximate Entropy estimation (avg.±std) in MI patients with normal and reduced ejection fraction ($p<0.05$).

5. Acknowledgments

The author is very grateful to Dr Federico Lombardi and Dr. Stefano Guzzetti, Dept. of Internal Medicine, L. Sacco Hospital, University of Milan, Italy, for providing experimental data.

6. References

Abarbanel H.D.I., R. Brown, J.J. Sidorowich, L.S. Tsimring (1993) "The analysis of observed chaotic data in physical systems", *Rev. of Modern Physics*, 65, 4:1331-1392.

Beran, J. (1994) "Statistics for Long-Memory Processes" Chapman & Hall, New York.

Cerutti S., G. Carrault, P.J.M. Cluitmans, A. Kinie, T. Lipping, N. Nikolaidis, I. Pitas, M.G. Signorini (1996) "Nonlinear algorithms for processing biological systems" *Comp. Meth. & Progr. in Biomedicine*, 51:51-73.

Flandrin, P. (1989) "On the spectrum of fractional Brownian motions", *IEEE Trans. on Information theory*, 35, 1:197-199.

Guzzetti S., M.G. Signorini, C. Cogliati et al. (1996) "Deterministic chaos indices in heart rate variability of normal subjects and heart transplanted patients" *Cardiovasc. Res.* 31:441-446.

Higuchi T. (1988) "Approach to an irregular time series on the basis of the fractal theory" *Physica D* 31:277-283.

Higuchi T. (1990) "Relationship between the fractal dimension and the power law index for a time series: a numerical investigation" *Physica D* 46:254-264.

Kamath M.V. and E.L. Fallen (1993) "Power spectral analysis of HRV: a non-invasive signature of cardiac autonomic functions" *Critical Rev. in Biom. Eng.* 21, 3:245-311.

Lombardi F., G. Sandrone, A. Mortara, D. Torzillo, M.T. La Rovere, M.G. Signorini, S. Cerutti, A. Malliani (1996) "Linear and nonlinear dynamics of heart rate variability after acute myocardial infarction with normal and reduced left ventricular ejection fraction" *Am. J. Cardiol.* 77:1283-1288.

Mainardi L.T., A. Yli-Hankala, I. Korhonen, M.G. Signorini et al. (1997) "Monitoring the Autonomic Nervous System in the ICU Through Cardiovascular Variability Signals" *IEEE-EMB Mag.* 16, 6:64-75.

Mandelbrot B.B. (1985) "Self-Affine Fractals and Fractal Dimension" *Physica Scripta*, 32:257-260.

Mandelbrot B.B., J.W. Van Ness (1968) "Fractional Brownian motions, fractional noises and applications" *SIAM Rev.*, 10, 4:422-437.

Osborne A.R., A. Provenzale (1989) "Finite correlation dimension for stochastic systems with power-law spectra" *Physica D* 35:357-381.

Pagani M., F. Lombardi, S.Guzzetti et al. (1986) "Power spectral analysis of a beat-to-beat heart rate and blood pressure variability as a possible marker of sympatho-vagal interaction in man and conscious dog" *Circ.Res.* 59:178-193.

Parker T.S., L.O. Chua (1987) "Chaos: a tutorial for Engineers" *Proc. of the IEEE* 75:982-1008.

Pincus S.M. (1995) "Approximate entropy (ApEn) as a complexity measure" *Chaos* 5:110-117.

Taqqu M.S., V. Teverovsky, W. Willinger (1996) "Estimators for long-range dependence: an empirical study" *Fractals*, 4, 3:785-798.

Turcott R.G., M.C. Teich (1996) "Fractal character of the electrocardiogram: distinguishing heart failure and normal patients" *Annals of Biomed. Eng.*, 24:269-293.

West B.J. (1990) *"Fractal physiology and chaos in medicine"* World Scientific.

ON THE CHARACTERIZATION OF ATRIAL FIBRILLATION USING WAVELETS AND CORRELATION DIMENSION

ALDO CASALEGGIO
Istituto Circuiti Elettronici, CNR, Via De Marini 6, Genova, I-16149, Italy.
E-mail: casaleggio@ice.ge.cnr.it

and

ROBERTO FIORAVANTI
Istituto Cibernetica e Biofisica, CNR, Via De Marini 6, Genova, I-16149, Italy.

ABSTRACT

This paper proposes a study on a combined use of the wavelets and the correlation dimension (D2) to obtain information on complex dynamics from the analysis of an observable of the investigated system. The processing consists of a filtering step (by means of wavelets decomposition of the signal) and a dimensional analysis step (by means of the correlation dimension estimation). The investigated dynamic are heart activities during normal sinus rhythm and atrial fibrillation (AF) through the analysis of electrocardiograms (ECG) obtained from MIT-BIH Arrhythmia Data Base. The study investigates the possibility to evidence the presence of low-dimensional sub-dynamics from the whole time series. Results show that, in the considered cases, there is a significant difference between the dynamics underlying certain detail levels derived from AF tracings and the same detail levels obtained from NSR tracings. This suggest a possible use of combined wavelets decomposition plus reconstruction and correlation dimension estimation to get a deeper insight in activities underlying cardiac arrhythmia.

This study is dedicated to the use of wavelets as tools for filtering signals in order to distinguish between the various sub-system involved in a dynamic. Dynamic is monitored by means of the correlation dimension and it is considered to be significant if it is possible to distinguish a clear linear scaling regions from the correlation integral plot. Correlation dimension computation is not the topic of this paper and it is discussed elsewhere (Ott *et al.*, 1994; Grassberger & Procaccia, 1983; Casaleggio *et al.*, 1995). Similarly wavelets analysis is discussed in (Daubechies, 1988; Daubechies, 1993; Ramchandran *et al.*, 1996). In this work wavelets decomposition and reconstruction has been done using the commercially available *Wavelets Toolbox of MATLAB* (Misiti *et al.*, 1996).

The major point of this paper is its focus on the possibility that wavelets themselves could influence the estimation of the correlation dimension. Such a situation was considered in (Badii & Politi, 1989) as a possible consequence of a filtering activity. To investigate the influence of wavelets on the correlation dimension estimation, we simply considered different kind of wavelets and we examined the differences in the obtained D2 and scaling regions estimates.

The processing consists of (i) to decompose the signal using wavelets; (ii) to reconstruct the new signal as the detail level which better represents atrial activity (under the condition that the same detail level is used for the various considered kind of wavelets); (iii) to compute the correlation dimension of such a build time series and (iv), to verify if there is a scaling region and, if that is the case, the correlation dimension value is used to characterize the system. As an ansatz, we assume that if the scaling region is convincingly visible, there is an underlying regularity in the time series since a low-dimensional attractor can be seen and quantified, and thus we accepted to consider as a valid analysis the correlation dimension without a-priori tests on the signal stationarity. Just we propose to accept the a-posteriori presence of a linear scaling region as an indication of an underlying stationary dynamic.

The study has been carried on ECG recordings of the MIT-BIH Arrhythmia Database (MIT-BIH Directory, 1982) sampled at 360 Hz; we analyzed mit113, mit115, mit117 whose predominant rhythm is normal sinus rhythm (NSR) and two pieces of mit203 whose predominant rhythm is atrial fibrillation (AF). D2 estimate of the original signals have been done on 20000 points, while reconstructed signals obtained from the fifth detail level after wavelets decomposition last 4096 points).

We remind that in a previous work (Casaleggio *et al.*, 1996) wavelets have been used to filter possible dynamics from the various band of a signal and we had a first preliminary result in which we saw an apparently clear scaling region just before the starting of ventricular fibrillation from the analysis of mit207 in the frequency band between 45 and 90 Hz. The principle is similar. In this case we simply consider details levels instead of reconstructed signals of specific bandwidth.

It is known that more than one dynamics underlies AF: in fact, atrial activity is continuously active, while the ventricles follows a more regular dynamics owing the filtering effect of the atrio-ventricular junction. Both dynamics (atrial and ventricular) have frequency components within the same band, then their separation based on spectral methods can not be usefully applied.

The filtering phase consist of wavelets decomposition and reconstruction from the fifth detail level of the signal decomposed by using Haar, Daubechies 2 and Symlet 5 wavelets. In such a way, the three filters are intrinsically strongly different. This allows us (i) to make some preliminary, although useful, consideration about the influence of the considered wavelets on D2 estimation and (ii) to draw some conclusion about underlying dynamics when the D2 estimates about the same ECG tracings are similar with the different filters. In this case it is more reasonable to assume that the measured D2 is to be attributed to the underlying dynamics rather than to the filtering process.

An example of the obtained original and filtered signals is presented in Fig.1 where the case of mit203 is investigated in detail. Detail levels N. 5 are reported in the figure for all the considered decomposed signals on the bases of Sym 5, Daubechies 2 and Haar wavelets. The expected result is the extraction of the baseline (ideally elimination of QRS and waves).

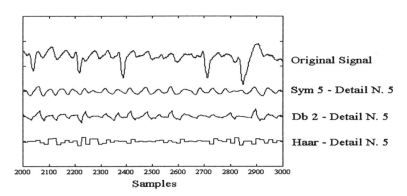

Figure 1: Example of filtering a piece of mit203 using the different wavelets.

A quantitative presentation of the obtained D2 estimates is reported in Table 1.

Table 1: D2 estimates obtained from the original signal and from the corresponding filtered signals (detail level 5 of the various wavelet).

file	rhythm	Original D2	α	Daubechies D2	α	Symlet D2	α	Haar D2	α
113	NSR	0.7 ± .03	.01	1.1 ± .07	.05	1.5 ± .06	.04	0.4 ± .03	.02
115	NSR	2.9 ± .26	.16	0.7 ± .05	.03	0.8 ± .08	.05	0.8 ± .06	.04
117	NSR	1.2 ± .02	.01	1.1 ± .02	.01	0.7 ± .03	.02	0.8 ± .02	.01
203a	AF	2.4 ± .24	.15	2.6 ± .13	.08	2.6 ± .12	.07	1.2 ± .03	.01
203b	AF	5.5 ± .34	.21	2.2 ± .04	.01	2.9 ± .20	.12	1.7 ± .07	.04

We note some differences between the D2 estimation of the original signal and those obtained by Symlet and Daubechies decomposition and reconstruction. Although results are preliminary because of the limited number of considered ECG signals and wavelets have been used in a very rough way, and this was required because we are interested in an extremely simplified analysis, there is evidence that the baseline dynamics is more constant or periodic in NSR than in AF since we obtained D2 values around 1 or lower in the case of NSR and greater than 2 in the case of AF. The α value of the Table is discussed in (Casaleggio et al., 1995). It indicates the degree of reliability of the D2 estimation. Roughly the smaller is α, the greater is D2 reliability.

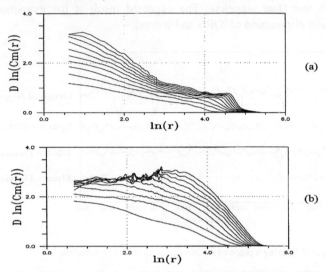

Figure 2: Log-log plot obtained from detail 5 of the analyzed mit117 and mit203 recordings. It is clear the presence of the scaling region.

The scaling regions were manually searched in the log-log plot. As examples, in Fig.2 we report the case of the derivative of the log-log plot of the filtered mit117 (Fig.2a) and mit203a (Fig.2b). We note the presence of acceptable scaling regions in both cases. We propose to consider such observation as an indication of an underlying process possibly involved in the cardiac activity pertaining the considered detail level. Similar log-log plot are obtained using Daubechies or Symlet wavelets, while a different pattern is observed from the analysis of the one obtained using Haar wavelet.

A comment regarding the differences in D2 estimation obtained from the filtered signals using Symlet or Daubechies wavelets and those obtained using Haar wavelet, is necessary. We note that Haar decomposition introduces many constant intervals in the produced time series (see Fig.1), and we remind that in a previous paper (Casaleggio, 1992) it was observed that constant intervals within a time series induce a strong underestimate of the correlation dimension. This is due to the fact that several points are mapped in a single point within the reconstructed phase space. Thus we observe that Haar wavelet is not useful for us.

In conclusion we investigated the combined use of wavelets decomposition and correlation dimension for the study of cardiac arrhythmia since it allows not invasive monitoring of dynamics underlying atrial activity from ECG recording.

We found some preliminary evidence of strong differences between the filtered ECG signal in the case of normal sinus rhythm and atrial fibrillation based on the D2 estimate of the detail level which better represent the baseline of ECG tracings. At a more speculative level, this might allow to monitor and quantify the action of some therapeutically chemical during time, or to verify whether different patients react in a similar way to some medical agents. Moreover, we get preliminary evidence that it is possible to use signal decomposition by means of wavelets in order to enhance the ability to locate low-dimensional activities in cardiac arrhythmia.

References

Badii, R. and A. Politi (1989) "On the Fractal Dimension of Filtered Chaotic Signals" in: *Dimension and Entropies in Chaotic Systems,* G. Mayer-Kress ed., Berlin: Springer, pp. 67-73.

Casaleggio, A. (1992) "Effect of Constant Intervals of a Time Series on the Computation of the Correlation Dimension" in: *Comp. in Cardiology 1992,* A.Murray and R.Arzbaecker eds, Los Alamitos: IEEE Comp. Soc. Press, pp.455-458.

Casaleggio, A., A. Corana and S. Ridella (1995) "Correlation Dimension Estimation from Electrocardiograms" *Chaos Solitons and Fractals* , 5:713-726.

Casaleggio, A., B. Gramatikov and N.V. Thakor (1996) "On the Use of Wavelets to Separate Dynamical Activities Embedded in a Time Series" in: *Comp. in Cardiology 1996,* A.Murray and R.Arzbaecker eds, Los Alamitos: IEEE Comp. Soc. Press, pp.181-184.

Daubechies, I., (1988) "Orthogonal Bases of Compactly Supported Wavelets." *Commun. on Pure and Applied Math.,* 41:909-996.

Daubechies, I., (1993) "Orthogonal Bases of Compactly Supported Wavelets ii Variations on a Theme." *SIAM J. on Math. Analysis,* 24:499-519.

Grassberger, P. and I. Procaccia (1983) "Measuring the Strangeness of Strange Attractors" *Physica D,* 9:189-208.

Misiti, M., Y. Misiti, G. Oppenheim and J.M. Poggi (1996) *Wavelet Toolbox,* The MathWork Inc.

MIT-BIH Arrhythmia Database Directory (1982) - BMEC TR010; revised version August 1988.

Ott, E., T. Sauer and J.A. Yorke (1994) *Coping with Chaos,* Wiley, New York.

Ramchandran, K., M. Vetterli and C. Herley (1996) "Wavelets, Subband Coding, and Best Bases." *Proc. of IEEE,* 84:541-560.

DYNAMICAL SYSTEM MODELLING AND CONTROL OF A CARDIAC ARRHYTHMIA

T. NOMURA, Y. SAEKI and S. SATO

Graduate School of Engineering Science
Osaka University, Toyonaka 560 Osaka, Japan
E-mail : taishin@bpe.es.osaka-u.ac.jp

ABSTRACT

Reentrant tachycardia is modeled as periodic circular movement of an action potential on a two dimensional excitable medium with circular geometry. Clinical observations of the response of the heart to both single and periodic electric stimulation delivered during reentrant tachycardia, such as phase resetting, termination and entrainment, are reproduced by the model. We analyze these phenomena and their dependency on stimulus period and initial phase as well as position of the stimulation on the medium by numerical integration of the model equations and by one dimensional maps (Poincaré maps).

1. Introduction

Fluctuations of heart beat rhythm are caused by various factors such as physical activity, respiration and autonomic nerve activity. These are referred to as *arrhythmias*. Some tachycardias are believed to be caused by circular movement of an excitation wave. For example, the presence of an extra pathway between atrium and ventricle other than the normal AV-node or that of an inexcitable area such that an artery or valve orifice which provide anatomical and/or electrical holes (Bernstein and Frame, 1990, Wellens et al., 1995) can trigger an excitation that circulates persistently around a closed circuit in the cardiac tissue. This periodic circulation, referred to as *reentrant tachycardia*, governs the heart contraction. Preventing the occurrence of the tachycardia is of first importance. Once it occurs, however, it is important to terminate it since this arrhythmia can initiate more dangerous ones such as *ventricular fibrillation*.

One possible method to achieve this is to deliver pulsatile stimulation directly to the heart. Reentrant tachycardia displays four distinct types of response to such stimulation: (1) Phase of the oscillation is reset, (2) the tachycardia itself is terminated by a single stimulus or (3) by an appropriate number of appropriate equally spaced pulses. (4) The tachycardia is entrained by some periodic pulse stimuli. All these responses display phase dependence. Moreover, the location of the stimulating electrode in the heart influences the response (Josephson, 1993). Despite the importance of the problem to human health, the majority of clinical studies in this area are based on empirical evaluation of pacing algorithms.

In this paper, we study the reentrant wave propagation using a simple theoretical model. The model consists of coupled nonlinear ordinary differential equations. Each single cell is modeled by the well known excitable cell membrane model. They are coupled with cylindrical symmetry so that the coupled system can mimic a two dimensional excitable medium defined on the cylindrical surface. The choice of this geometry is suitable as it is common in cardiology to imagine a reentrant wave as a one dimensional ring (Berstein and Frame, 1990; Josephson, 1993), and the band geometry with small width is its natural extension (Nomura and Glass, 1996), although the geometry itself is rather artificial.

We explore the responses of the model exhibiting the reentrant wave to external current stimulation. The conditions to control and/or terminate the reentrant wave are analyzed. We will show that reentrant wave is either terminated, entrained, or switched to a different steady state (spiral-like wave) depending on the stimulus parameters.

In Sec. 2, we introduce the model excitable medium. The model equations have three qualitatively different stable solutions (attractors): (i) each element of the medium (single cell model) is in its resting state (stable equilibrium), (ii) the excitation wave propagates periodically on the medium, (iii) the excitation wave propagates aperiodically. (i) corresponds to the normal heart beat, that is, no reentrant wave, and (ii) and (iii) to the reentrant tachycardia. In Sec. 3, we show the system's dynamics when external inpulsive current stimuli are applied to the system at the state either (ii) or (iii). The stimulation corresponds to electric stimulation delivered directly to the heart through electrode. A single stimulation or equally spaced stimuli can induce the following transitions among the attractors: (ii)→(i), (ii)→(iii), (ii)→(iii)→(i), (ii)→(iii)→(ii) depending on timing (phase) of the first stimulation, inter-stimulus interval, and location of the stimulation on the medium. In Sec. 4, we show that these transitions can be predicted by using simple one-dimensional maps. In Sec. 5, the results are summarized.

2. The model

The well-known FitzHugh-Nagumo equations (FHN) are chosen as a single

Figure 1: Periodic reentrant wave (phase=0)
muscle cell model (FitzHugh, 1961):

$$\begin{cases} \dot{x} = c(x - x^3/3 - y + z) \equiv f(x,y), \\ \dot{y} = (x - by + a)/c \equiv g(x,y), \end{cases} \quad (1)$$

where x represents membrane potential, y the refractoriness. The parameters are set to $a = 0.7$, $b = 0.5$, $c = 3.0$, $z = -0.3$ so that the model cell behaves as excitable membrane.

FHN models are configured on $M \times N$ two dimensional grid points (i,j) ($i = 1, \cdots, M$, $j = 1, \cdots, N$). The cell located at the (i,j) grid point is referred to as the $k = [N(i-1) + j]$th cell. The coupled system can be written as:

$$\begin{cases} \dot{x}_k = f(x_k, y_k) + D_s \sum_{k'}(x_{k'} - x_k), \\ \dot{y}_k = g(x_k, y_k), \end{cases} \quad (2)$$

where (x_k, y_k) denotes the state of the k-th cell, and $\sum_{k'}$ denotes summation for the k-th cell over the eight neighboring cells. D_s represents the conductance of the electrical coupling corresponding to the low resistance intercellular channels between muscle cells called gap junction. The two vertical boundaries of the square are connected to one another, but the horizontal boundaries are not. Thus, the coupled system possesses cylindrical symmetry. Since the system is described by ordinary differential equations and therefore it has no spatial dimension, the model dynamics may still correspond to the dynamics of two dimensional excitable media defined on the cylindrical surface and we call the model the two dimensional excitable medium.

We numerically integrate eqs.(2) by using second order Euler scheme for ordinary differential equations ($\Delta t = 0.02$). An appropriate initial value for each cell can easily induce a reentrant excitation wave circulating along the cylinder surface periodically. As illustrated in Fig. 1, the excitation wave propagates and circulates from left to right direction. The time interval in which a large enough stimulus can induce an excitation other than the reentrant wave is referred to as excitable gap G_s. $R_s = G_s/T_s$ is the ratio of the excitable gap to the reentry period. We calculate a reentry period T_s by detecting the passage of membrane potential across a threshold at the recording site.

The values of M, N, D_s are chosen by simulating various sets of values systematically as follows: T_s and G_s are calculated for every choice of M, N and D_s. The reentrant period T_s increases proportionally to N (larger slope for smaller D_s) unless N is too small (~ 20), and it decreases exponentially as M increases. The ratio R_s is a square root-like function of N for $N > 20$, and larger R_s is obtained for smaller D_s for a fixed N. We referred to the data by Bernstein and Frame (1990) and chose the following experimental values: The length of the reentrant pathway, $L_e = 99$mm, the reentrant period $T_e = 450$ms, excitable gap $G_e = 200$ms ($R_e \sim 0.44$). Here we set $N = 65$, $D_s = 0.7$ in the model. When $M = 9$, we obtain $T_s \sim 25.0$tu and $R_s \sim 0.47$. Comparing these values with the experimental data, the size of a single cell model (the space units=1su) and the time unit=1tu of the simulated results can be interpreted in terms of physical dimension. In this case, we have the following correspondence: 1su= $L_e/N \sim 1.5$mm, 1tu= $T_e/T_s \sim 18$ms, and

$$D_s[\frac{su^2}{tu}] = D_s\frac{(L_e/N)^2}{(T_e/T_s)} = D^*[\frac{mm^2}{ms}] \sim 0.09\frac{mm^2}{ms} \sim 1.3\frac{cm^2}{s},$$

where D^* represents diffusion constant of the real heart tissue. A similar value of D^* is obtained for $M = 7$. We use $M = 9$ and $M = 7$ in the following simulations.

3. Response to a single stimulation

Since the periodic reentrant wave (Fig. 1) is a limit cycle solution of the coupled ordinary differential equations (2), we can define the phase ϕ describing the system's state along the cycle. The state when the reentrant wave with period T_s passes through a recording site is defined to be $\phi = 0$. The phase of the state t tu ($0 \leq t \leq T_s$) after the phase 0 state is defined to be $\phi = t/T_s$. When the wave passes through the recording site again ($t = T_s$), the phase is reset to 0 ($\phi = 1 = 0 \pmod 1$). As in cases of a single limit cycle oscillator, an inpulsive stimulation applied to the medium can induce phase delay or phase advance depending on the stimulus phase, where "the stimulus phase" represents the phase of the reentrant wave when the stimulus is applied. We also consider the influence of the stimulus location on the dynamics. Because of the cylindrical symmetry of the medium, we always apply the stimulations on the 30-th line ($i = 30$), and vary the column j systematically. We set the recording site at the same position as the stimulus location.

Figures 2 to 5 are the typical responses of the model to single stimulation. Figures 2, 3 and 4 are for the model with $M = 7$, and Fig. 5 is for $M = 9$. In each case, the stimulus location is set to $(i, j) = (30, 2)$.

Figure 2 illustrates a phase resetting by a stimulation (the stimulus phase $\phi \sim 0.31$). The excitation induced by the stimulus applied at the center of the excitable gap propagates in both directions (Fig. 2(a) and (b)). The wave propagating to the left eventually collides with the reentrant wave. The collision

Figure 2: Phase reset of the reentrant wave by a single stimulation

Figure 3: Termination of the reentrant wave by a single stimulation

leads to the termination of both the reentrant and the induced waves (Fig. 2(c)). The wave propagating right remains and establishes a reset reentrant wave. In Fig. 3, the excitation induced by the stimulus (the stimulus phase $\phi \sim 0.276$) propagates only left (Fig. 3(a) and (b)), and collides with the reentrant wave (Fig. 3(c)), leading to the termination of all the excitation wave.

In Fig. 4 and 5, the single stimulation induces a spiral-like wave. In Fig. 4 (the stimulus phase $\phi \sim 0.24$), the core of the spiral progressively drifts to the left-lower of the medium (Fig. 4(a) and (b)). Eventually it reaches the boundary of the medium, where it turns into two left and rightward propagating plane waves. These two waves collide with each other, and disappear (Fig. 4(c) and (d)). Note that the spiral-like wave survives several times of the reentrant period T_s before reaching the state depicted in Fig. 4(c). In Fig. 5, (the stimulus phase $\phi \sim 0.28$), the core of the spiral-like wave drifts leftward persistently, and the reentrant wave never disappears. In this case, the inter-action potential interval at the recording site is quasi-periodic.

Figures 6(a) and (b) summarize responses of the medium to single stimula-

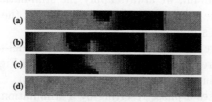

Figure 4: Transient spiral-like wave induced by a single stimulation

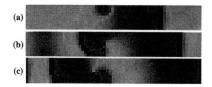

Figure 5: Spiral-like wave induced by a single stimulation

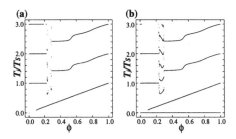

Figure 6: The phase resetting curves: (a)$M = 7$, (b)$M = 9$.

tions for $M = 7$ and $M = 9$, respectively. The abscissa is the stimulus phase, and the ordinate is the time, normalized by the reentrant period T_s, from the detection time of a reference action potential at the recording site (the zero of the ordinate). For each stimulus phase, we plot dots when the j-th action potential is detected at time T_j (the phase resetting curve). When the stimulus phase is small, the responce of the medium is negligible and the successive action potentials are detected almost every $T_s$ms after the reference (flat portions of Fig. 6(a) and (b)). When the stimulus phase is larger than $\phi \sim 0.29$, in both (a) and (b), arrivals of the successive action potentials are advanced in comparison with the cases without the stimulation.

When $0.226 < \phi < 0.289$ in (a) and $0.231 < \phi < 0.294$ in (b), the phase resetting curves are complicated. We refer to these intervals as "the critical windows." In (a), there is no plot in the critical window. When the stimulus phase is in the window or the border of it, the reentrant wave is terminated almost immediately or via transient spiral-like wave, respectively. In (b), the periodic reentrant is altered to the spiral-like wave when the stimulus phase is in the critical window. In other words, the stimulus within the critical window moves the state of the system out of the basin of attraction of the limit cycle (the periodic reentrant wave).

When the stimulus phase is ϕ, the system's phase is shifted by $\Delta\phi = T_j - jT_s$ (for large enough j). $\Delta\phi > 0$ and $\Delta\phi < 0$ represent phase delay and advance, respectively. Then the relation between the phase just before and after the stimulation is given by $g(\phi)$ as follows:

$$g(\phi) = g_j(\phi) = (\phi - T_j/T_s) \mod 1 \quad (j \to \infty). \tag{3}$$

Note that the phase just after the stimulation has not been defined, since just after the stimulation, the state has not been recovered to the limit cycle. The usage of the term phase for this state can be justified by using the concept "isochronal phase" defined for a limit cycle oscillator (Winfree, 1987, Kawato, 1981, Nomura et al, 1994). The graph of g is called the Phase Transition Curve (PTC). In the subsequent section, we analyze the dynamics of the model when it is stimulated periodically using one dimensional discrete maps derived from the PTC.

4. Responses to periodic stimulation

In this section, we apply a finite number of equally spaced impulsive stimulations to the system, and analyze the system's dynamics. The system's behavior can be described approximately by an appropriate one dimensional map model derived from the PTC.

When the n-th stimulation is applied at the phase ϕ_n, the system's phase moves to $g(\phi_n)$ approximately. Let us denote T and w the inter-stimulus interval and pulse width, respectively. Then, since the $(n + 1)$-th stimulation will be applied $(T - w)$ ms after the n-th stimulation, ϕ_{n+1} is written in the following form using the one dimensional map f:

$$\phi_{n+1} = g(\phi_n) + \tau - d \pmod 1 \equiv f(\phi_n), \tag{4}$$

where $\tau = T/T_s$ and $d = w/T_s$. The map f approximates the Poincaré map of the periodically perturbed limit cycle oscillator. That is, given ϕ_1, successive phase sequences $\{\phi_n\}$ can be obtained using f iteratively. The dynamics of the medium can be described by those of the one dimensional map. A p-periodic sequence of the map, for example, corresponds to a p periodic entrainment between the reentrant wave and the periodic stimulation. If the stimulus falls in the critical window defined above, the iteration of the map is terminated, and this corresponds to the transition of the system's state from the periodic reentrant to either termination or the spiral-like wave.

Figure 7 examplifies the dynamics of the map when the ratio between the stimulus period and the period of the reentry is 0.3. The first stimulation is applied at $\phi_0 = 0.9$, and induces the phase shift about 0.3. The second stimulation is delivered at $\phi_1 \sim 0.28$ for which PTC is not defined, and leads to the termination of the reentrant wave. In this case, the second stimulation falls

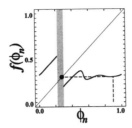

Figure 7: Dynamics of the one dimensional map

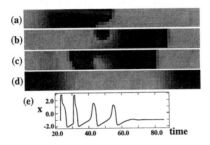

Figure 8: A transient spiral-like wave induced by two stimulations.

at the border between the window and the phase resetting portion (Fig. 6(a), $M = 7$).

Figure 8 illustrates the dynamics of the medium which corresponds to the dynamics of the map shown in Fig. 7. The inter-stimuli interval is 30 % of the period of the reentry, and the phase of the first stimulation is 0.9. Figure 8(a) shows the system's state just after the first stimulation, (b) just after the second one, (c) induced transient spiral-like wave, and (d) termination of the reentrant wave after some period. In Fig. 8(e), the corresponding action potential waveform at the recording site is displayed. The dynamics coincides well with the prediction of Fig. 7.

5. Conclusion

Reentrant excitation wave can be supported by an excitable medium with cylindrical geometry. The reentrant wave will be either reset, terminated or switched to the spiral like wave by a single stimulus depending on the stimulus phase and its location. We have demonstrated that one can predict the dynamics of the medium when it is stimulated by multiple equally spaced stimuli once the phase resetting curve describing the effect of a single stimulus and its stimulus phase dependence is known. That is, multiple stimuli will either terminate the reentry, entrain the reentry (figures not shown for entrainments), or switch the periodic reentry to aperiodic spiral-like wave, depending on the inital phase of the stimulus, the interval between the successive stimuli and the number of the stimuli.

Acknowledgment

The authors would like to thank Leon Glass (McGill Univ.) and K. Pakdaman (Osaka) for fruitful discussions. This research has been partially supported by Grants-in-Aid for Scientific Research Nos (2)09268224 and (c)(2)09680 853, The Japanese Ministry of Education, Science and Culture.

References

Bernstein RC. and L.H. Frame (1990) "Ventricular Reentry around a fixed barrier. Resetting with advancement in a in vitro model", *Circulation* 81: 267-280

Wellens H.J., J.L. Smeets, A.P. Gorgels, J. Farre (1995) "Wolf-Purkinson-White Syndrome", in *Cardiac Arrhythmias 3rd edition*, Mandel W.J. ed., JB Lippincott Company

Josephson M.E. (1993) *Clinical Cardiac Electrophysiology. 2nd edition*, Philadelphia: Lea and Febiger

Nomura T. and L. Glass (1996) "Entrainment and Termination of Reentrant Wave Propagation in a Periodically Stimulated Ring of Excitable Media", *Physical Review E* 53-6: 6353-6360

FitzHugh R. (1961) "Impulses and physiological states in theoretical models of nervemembrane", *Biophy J* 1:445-466

Winfree AT. (1987) *When Time Breaks Down*, Princeton University Press

Kawato M. (1981) "Transient and steady state phase response curve of limit cycle oscillators", *J. Math. Biol.* 12:13-30, 1981

Nomura T., S. Sato, S. Doi, J.P. Segundo, M. Stiber (1994) "Global bifurcation structure of a Bonhoeffer van der Pol oscillator driven by periodic pulse trains", *Biological Cybernetics* 72:55-67

INVESTIGATION OF PASSIVE ELECTRICAL PROPERTIES IN THE RIGHT ATRIUM

ARTŪRAS GRIGALIŪNAS
Kaunas Medical Academy, Eiveniu 4, Kaunas, 3007, Lithuania
E-mail: agrigal@gim.ktu.lt

and

ROMUALDAS VETEIKIS
Institute for Biomedical Research, Kaunas Medical Academy, Eiveniu 4, Kaunas, 3007, Lithuania

ABSTRACT

Measurements of electrotonic potential distribution in the right atrium of the rabbit heart were made. A KCl-perfused suction electrode was used as the current source. Space constants along (λ_x) and across (λ_y) the fibres and the time ($t_{1/2}$), during which the electrotonic potential reached half the stationary amplitude were calculated. Experimental results were analyzed by using a two-dimensional anisotropic homogeneous RC-medium. The dependence of $t_{1/2}$ upon the distance (r) between the current source and the measurement point was studied. The obtained inadequacy from a linear dependence, already reported, can be explained as dependence on 1) dimensions of the current source; 2) anisotropy of the passive electrical properties in the right atrium; 3) inhomogeneity of the cardiac tissue.

1. Introduction

Research into the passive electric properties of the cardiac tissue is important for the analysis of the excitation spread processes, mechanisms of the arrhythmia genesis and the generation of ectopic foci. In order to estimate the distribution of the electrotonic potential in anisotropic syncytial media, current can be injected intracellulary by means of a circle-shaped suction electrode. When suction electrode is used (Adomonis *et al.*, 1983), certain advantages and limitations of the technique are entailed, if compared with the use of a microelectrode or the chamber technique of current injection. The electrotonic potential abruptly decreases, while spreading away from the microelectrode used as the current source (Woodbury & Crill, 1961). That is why accuracy of the measurement decreases. Some authors applied the chamber technique, alongside with a microelectrode as the current source, in thin cylindrical tissue pieces excised from the myocardium (Nishimura, 1988). However, in cylinder-shaped pieces excised from tissues lacking a definite cell orientation, the cylinder axis may not coincide with the direction of the optimal electrotonic spread. If a high electrotonic anisotropy exists, the value of the electrotonic space constant measured by means of the chamber technique of current application is highly dependent on the direction of excision. With a suction electrode used as the current source, the above drawbacks are avoided. However, other problems arise, related to the selection of models for the electrotonic potential distribution, and to the description of a suction electrode as the current source.

2. Materials and methods

Experiments where made following standard methods of electrophysiological research in vitro. The isolated preparation of the right atrium was mounted in a tissue bath, perfused with a Tyrode solution. KCl-perfused suction electrode was used to inject current pulses with an amplitude of 10^{-5}-10^{-6}A into the endocardial side of the preparation. The electrotonic potential was recorded by means of a microelectrode in different points of endocardium. Space constants along (λ_x) and across (λ_y) the optimal electrotonic spread where calculated referring to the analysis of the electrotonic potential amplitude decay (Bukauskas et al., 1982). In order to evaluate the time constant of the electrogenic membrane (τ_m), the dependence of time ($t_{1/2}$) during which the electrotonic potential reaches half the stationary amplitude upon the distance between the current source and the registration point was explored.

3. Results and Discussion

The distribution of electrotonic potential $V=V(x,y)$ in an infinite plane anisotropic cell of thickness h is described as

$$\lambda_x^2 \frac{\partial^2 V}{\partial x^2} + \lambda_y^2 \frac{\partial^2 V}{\partial y^2} = V + \tau_m \frac{\partial V}{\partial t} \quad (1)$$

where $\lambda_x = (R_m h / 2\rho_x)^{1/2}$, $\lambda_y = (R_m h / 2\rho_y)^{1/2}$, $\tau_m = R_m C_m$ (R_m, C_m are specific resistance and capacitance of the electrogenic membrane; ρ_x and ρ_y are specific resistances of the intracellular medium) (Veteikis, 1991). In a normalized system of coordinates ($X = x / \lambda_x$, $Y = y / \lambda_y$) solution of the Eq. 1 for rectangular current step $I(t) = I_o$, applied at origin is:

$$V(R,T) = \frac{\rho I_o}{4\pi h}\left\{K_o(R) + \int_0^{\ln\frac{2T}{R}} \exp(-R\cosh w)dw\right\} \quad (2)$$

where $R = (x^2 / \lambda_x^2 + y^2 / \lambda_y^2)^{1/2}$, $\rho = (\rho_x \rho_y)^{1/2}$, $T = t / \tau_m$ and $K_o(R)$ is the McDonald function. A linear dependence between $t_{1/2}$ and R was demonstrated for one-dimensional, two-dimensional isotropic (Jack *et al.*, 1975, pp. 35-37) and two-dimensional anisotropic (Veteikis, 1991) media.

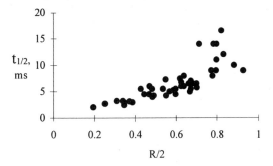

Fig. 1. Experimental dependence of half-time $t_{1/2}$ upon the distance R/2.

Our experimental results displayed in Fig. 1 suggest that the dependence between $t_{1/2}$ and R is not linear. To explain these results, we applied the mathematical model of two-dimensional anisotropic RC-medium, with a circumference-shaped current source. In accordance with the superposition principle well-known in electrostatics, the source of current was approximated by M equivalent point sources (X_i, Y_i) placed on the perimeter of the current source (Bukauskas *et al.*, 1991). The response V_i, produced by one point current source can be calculated by using Eq. 1. The total amplitude of electrotonic potential $V(X_i, Y_i)$ at the measurement point (X, Y) is equal to the sum of responses produced by all the point sources located on the perimeter of the circumference:

$$V(X,Y,T) = \sum_{i=1}^{M} V_i(R_i, T) \qquad (3)$$

Calculations of the half-time dependence on distance (R) between the measurement

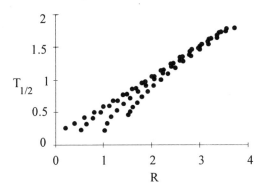

Fig. 2. Theoretical dependence of $T_{1/2}$ on R, when $\lambda_x=1$mm, $\lambda_y=0.1$mm, the radius of suction electrode $r_0=0.1$mm

point and the center of the circle-shaped current source, based on the above mentioned model (Fig. 2), show, that linear dependence can be observed only at long distances, where $r > 3\lambda$. At short distances, the nonlinear dependence comes out. Moreover, the sensitiveness to the y coordinate appears.

Our experimental measurements where made at a less than 2λ distances from the current source. Close to the current source the dependence of half-time on the distance declines from linear. Thus we can maintain that the model of two-dimensional anisotropic homogeneous medium with a point current source involves a computational error (that must be considered) if the current source is of finite value during the experiment. No doubt, the distribution of points as shown in Fig. 1 must have also been influented by inhomogeneity of cardiac tissue. Yet the model of two-dimensional anisotropic RC medium with a circle-shaped current source (not a point) has been helpful in proving the importance of the size of the current source.

References

Adomonis V., J. Bredikis and F. Bukauskas (1983) "The Suction electrode with inner perfusion" *Sechenov Physiol. J. USSR* 59: 272-274.

Bukauskas F., A. Bytautas, A. Gutman and R. Veteikis (1991) "Simulation of passive electric properties in the two- and three- dimensional anisotropic syncytial medium", in: *Intercellular Communication*, F. Bukauskas, ed., Manchester, New York: Manchester Univ. Press, pp. 203-217.

Bukauskas F., A. Gutman, K. Kisunas and R. Veteikis (1982) "Electrical cell coupling in rabbit sinoatrial node and atrium: experimental and theoretical evaluation" in: *Cardiac Rate and Rhythm. Physiological, morphological and developmental aspects*. L.N. Bouman, H.J. Jongsma, eds, Hague, Boston, London: Martinus Nijhof publ., pp. 195-216.

Jack J.J., D. Noble, R.W. Tsien (1975) "Electric Current Flow in Excitable Cells" Oxford: Clarendon Press.

Nishimura M., Y. Habuchi, S. Hiromasa et al. (1988) "Ionic basis of depressed automaticity and conduction by acetylcholine in rabbit AV node" *Am. J. Physiol.* 255: H7-H14.

Veteikis R. (1991) "Estimation of the membrane time constant in the two-dimensional RC-medium" *Biofizika* 36: 537-540.

Woodbury J.W. and W.E. Crill (1961) "On the problem of impulse conduction in the atrium", in: *Nervous Inhibition*, Florey E., ed.. Oxford: Pergamon Press, pp. 1124-1135.

ON THE STOCHASTICITY OF NEURONAL FIRING PATTERNS: THE LOCAL STABILITY OF LIMIT CYCLES

FATHEI ALI and MICHAEL MENZINGER
Department of Chemistry, University of Toronto
Toronto, Ont. M5S 3H6, Canada
E-mail: menzinger@chemistry.utoronto.ca

ABSTRACT

Limit cycles are usually made up of locally stable and unstable segments during which perturbations decay and grow, respectively. Depending on the strength of the local instability, diffusion-like dynamics, amplification and even expansion of phase-volumes may occur that play key roles in the noise-response of neural systems. The van der Pol oscillator is presented as an illustrative example.

1. Introduction

The origin and the function of irregularities in neuronal firing patterns are issues of considerable current interest (Anishchenko, 1995; Sauer, 1997). On the simplest, stochastic level, the problem may be reduced to the response to external noise of a limit cycle (LC) or a cycle arising from perturbations from an excitable state (Pikovsky & Kurths, 1997). A commonly held view is that a stable limit cycle is a closed trajectory to which neighboring trajectories are attracted everywhere, i.e. the cycle is locally stable everywhere. More realistically, the local rate of decay of perturbations to which we refer also as local stability, generally varies in phase space. It usually happens that a globally stable LC is composed of several locally stable, attracting and of locally unstable, repelling segments in which perturbations decay or are amplified, respectively, as illustrated by fig.1. Accordingly, the global stability of LCs arises only on average through the competition of local stability and instability (LI).

The local instability of strange attractors has been studied extensively (Benzi & Carnavale, 1989; Nese, 1989; Abarbanel, 1991; Jezewski, 1993) since it is a necessary condition for chaotic dynamics. However, LI persists in two-variable systems where chaos cannot occur and where the only alternative is the locally unstable LC illustrated by fig.1 and the corresponding local amplification of external noise. This was studied for the van der Pol oscillator (Treutlein & Schulten, 1986;

Kurrer & Schulten, 1991) by computing the stationary probability distribution function using the Fokker-Planck equation. The aim of this paper is to review the notion of LI and to illustrate it using the van der Pol model, and to indicate by simulation the consequences of local stability and instability. A more complete version will be published elsewhere (Ali & Menzinger, 1997.a).

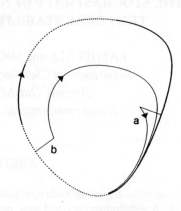

Fig.1: Perturbations applied during unstable and stable phases evolve differently.

2. Measures of Local Stability

Consider the dynamical system

$$\frac{d}{dt}x(t) = F[x(t)] \tag{1}$$

where F defines a flow in R^n. Let $x_o(t)$ be a non-stationary reference solution (e.g. a limit cycle) and $x(t)=x_o(t)+\delta x(t)$ a perturbed trajectory. The initial perturbation $\delta x(t)$ is small enough so that its evolution is described by the linearized equation:

$$\frac{d}{dt}\delta x = A\delta x \quad \text{where} \quad a_{ij}(t) = \left(\frac{\delta F_i}{\delta x_j}\right)_{x_o(t)} \tag{2}$$

In some regions of phase space, perturbations decay and in others they grow. The notion of local stability and instability associated with this decay and growth gains quantitative meaning when viewed from different frames of reference in which $\delta x(t)$ is projected. Different measures of local stability are associated with these frames. We only enumerate some of them since a more detailed discussion is in the full paper (Ali & Menzinger, 1997.a).

a. The natural measures of local stability are the eigenvalues λ_i of **A**, i.e. the decay rates along the eigenvectors. They have the drawbacks of frequently being complex and the directions of the eigenvalues not being related to $x_o(t)$ in a physically meaningful way.

b. Projection of $\delta x(t)$ into the frame of phase variables x_i, (i=1...n) reproduces their fluctuations. The decay rates are given by the diagonal and off-diagonal elements of the stability matrix **A**.

c. Another meaningful reference frame are the direction tangential to the flow and those contained in the manifold transverse to the flow. This rotating reference system is adopted here for illustrations. The transformation of the original equation (1) from the fixed basis **x** to the rotating, orthogonal basis **x**' is achieved by the unitary rotation matrix **U**, which transforms the perturbation $\delta \mathbf{x}$ vector into $\delta \mathbf{x}'$. Hence

$$\frac{d}{dt}(\delta x') = \frac{d}{dt}(U \delta x) = B \delta x' \quad \text{where}$$

$$B = UAU^{-1} + \frac{dU}{dt} U^{-1} \tag{3}$$

The elements λ_{ij} of the stability matrix **B** in the rotating frame were evaluated analytically for a two-variable system (Ali & Menzinger, 1997.a). In two dimensions λ_{nn} and λ_{tt} are the rates of convergence normal and tangential to the trajectory. The latter averages to zero over a closed orbit, and the phase average of the former is the Lyapunov exponent. The directions (t,n) are partly uncoupled since $\lambda_{nt}=0$, i.e. a tangential perturbation remains tangential., while $\lambda_{tn} \neq 0$, i.e. a perpendicular perturbation couples to the tangential motion and leads to a resetting of the phase. The divergence of the flow div**F**=tr**A**=tr**B**=$\lambda_{nn}+\lambda_{tt}$ is the rate of contraction (or expansion) of volumes in phase space.

3. Numerical Results

These concepts are illustrated with the van der Pol oscillator :
$dx/dt=\varepsilon^{-1}(y+x-x^3/3)$, $dy/dt=-b_o+x$, where the ratio of time scales is controlled by ε and supercritical Hopf bifurcations occur at $b_o=\pm 1$. The elements λ_{nn}, λ_{tn} and λ_{tt} of the stability matrix **B** were evaluated from the analytical expressions (Ali & Menzinger, 1997) for the full phase space and along the computed limit cycle, for cycles far from (fig.2; $b_o=0.0$) and near the Hopf bifurcation (fig.3; $b_o=0.999$), both for a value of $\varepsilon =1.0$.

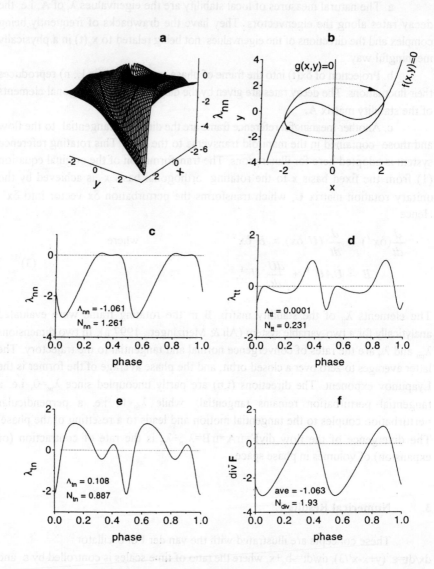

Fig.2: Local stability measures far from the Hopf bifurcation.

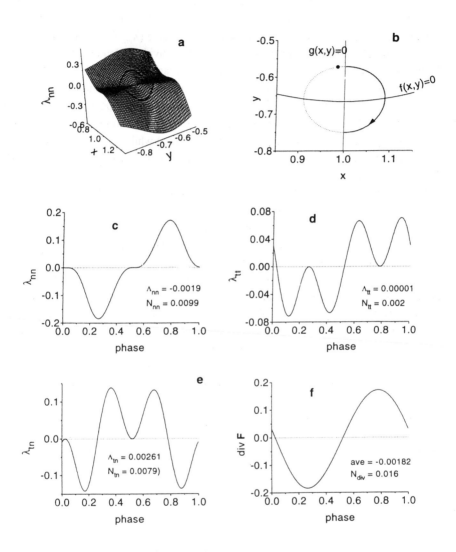

Fig.3: Local stability measures near the Hopf bifurcation.

Far from the bifurcation, fig.2a illustrates the dependence of the stability element λ_{mn} on the phase coordinates, and the trajectory of the LC is embedded in the stability landscape. The cycle with its stable and unstable ($\lambda_{nn}>0$), dotted portions is shown in fig.2b, together with the nullclines. The individual elements are shown as functions of oscillation phase (the solid dot in fig.2b marks zero phase) in figs. 2c,d,e, together with the corresponding phase averages Λ_{ij} and nonuniformities (variances) N_{ij}. During different phases, all three matrix elements assume positive values. The positive phase of the perpendicular element λ_{nn} is the least pronounced. As expected, the tangential element $\lambda_{\tau\tau}$ averages to zero ($\Lambda_{\tau\tau}=0.000$) while $\Lambda_{nn}= -1.061$ expresses the considerable global stability of the cycle. Apart from the phases where individual matrix elements take positive values, fig.2f shows that there are also phases of so-called 'strong local instability' where the divergence of the flow becomes positive. Here, phase space volumes increase before they are entrained by the globally contracting, dissipative flow. Additional computations at the same value of b_o have shown that decreasing the time-scale parameter ε only weakly enhances the local instability while strongly enhancing global stability.

Close to the Hopf bifurcation, fig.3, the stability landscape becomes flatter for all three matrix elements and for the flow divergence. The locally stable and unstable portions occur almost symmetrically, and the rapidly decreasing nonuniformities N_{ij} indicate that the local stability becomes uniformly neutral at the Hopf point, i.e. $\lambda_{ij}(t)=0.0$ for all elements. This neutral stability of the bifurcating Hopf cycle is the cause of its diffusive susceptibility to noise, as illustrated by fig.4b.

Finally, the effects of external dichotomous noise are illustrated for the two cases by the noisy limit cycles in fig.4. Far from bifurcation, the trajectories undergo diffusion-like motion in the unstable portion of phase space that occurs during the transition between 'slow' manifolds. After this they are rapidly attracted to the slow manifolds. Near the bifurcation, the cycle experiences its well-known diffusion-like drift, in accord with its neutral stability.

4. Remarks

The nonuniformity of the local stability measures (eigenvalues, rates of convergence; flow divergence) provides a fine structure of the system's response to perturbations. In particular, the unstable portions represent vulnerable windows through which the effect of external perturbations may be maximized. It may be of interest, for instance, to explore the relation of local stability and phase resetting. Of course, the response to noise depends on the spectral and statistical properties of the noise as well as on the system properties discussed here. Our analysis has exposed different phases of local stability: strongly damped and weakly or marginally damped phases which tend to undergo diffusion-like motion when perturbed randomly. For stronger instability, perturbations may be noticeably amplified, and beyond that the divergence of the flow may become positive, leading to a locally expanding flow in a dissipative system. This is reflected by the phase-dependent response of nonlinear oscillators to external perturbations (see e.g. fig.4 and Ali & Menzinger, 1997.b).

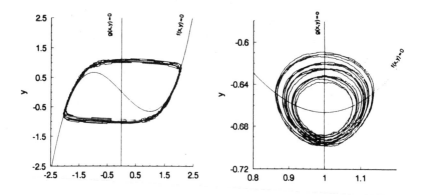

Fig.4: Noisy van der Pol cycles far from and near Hopf bifurcation.

The notion that the global stability of periodic attractors arises from the competition of locally stabilizing and destabilizing tendencies sheds light on the nature of bifurcations: they may occur smoothly (i.e. N=0) as in the Hopf case or abruptly, at finite nonuniformity, as in the saddle-node, supercritical Hopf or period doubling bifurcations (Ali & Menzinger, 1997.a). The concepts illustrated here by the van der Pol oscillator apply to nonlinear oscillators in general.

Acknowledgment

The work is supported by the NSERC of Canada.

References

Abarbanel, D.H.I., Brown, R. and Kennel, M.B. (1991), "Variation of a Lyapunov coefficient on a strange attractor", *J. Nonlin. Sci.* 1: 175.

Ali, F. and Menzinger, M. (1997.a), "On local stability of limit cycles", t.b.p.

Ali, F. and Menzinger, M. (1997.b), "Stirring Effects and Phase-Dependent Inhomogeneity in Chemical Oscillations: the B-Z Reaction in a CSTR", *J.Phys.Chem.* 101: 2304.

Anishchenko, V.S. (1995), "Dynamical Chaos: Models and Experiments", (World Scientific).

Benzi R. and Carnavale, G.F. (1989) "A possible measure of local predictability", *J. Atmos. Sci.* 46:3595.

Jeszewski, W. (1993) "Statistics of local expansion rates for chaotic systems", *Phys.Lett.* A183:63

Kurrer, C. and Schulten, K. (1991) "Effect of noise and perturbations on limit cycle systems" *Physica* D50:311.

Nese, J. (1989) "Quantifying local predictability in phase space", *Physica* D35:237.

Pikovsky, A.S. and Kurths, J. (1997) "Coherence resonance in a noise-driven excitable system", *Phys. Rev. Lett.* 78: 775.

Sauer, T. (1997), this volume.

Treutlein, K. and Schulten, K. (1986) "Noise-induced neural impulses", *Eur. Biophys. J.* 13: 355-59.

OSCILLATIONS IN CONTINUOUS-TIME GRADED-RESPONSE NEURAL NETWORKS WITH DELAY

K. PAKDAMAN[1,2], C.P. MALTA[3], C. GROTTA-RAGAZZO[4], O. ARINO[5]
and J.-F. VIBERT[2]

1) Department of Biophysical Engineering, Faculty of Engineering Science
Osaka University Toyonaka 560 Osaka, Japan
2) B3E, INSERM U 444, ISARS, Faculté de Médecine Saint-Antoine
27, rue Chaligny, 75571 Paris Cedex 12 FRANCE
3) Instituto de Física, Universidade de São Paulo
CP 66318, 05315-970 São Paulo, BRASIL
4) Instituto de Matemática e Estatística, Universidade de São Paulo
CP 66281, 05315-970 São Paulo, BRASIL
5) Laboratoire de Mathématiques Appliquées
Université de Pau et des Pays de l'Adour, IPRA, URA CNRS 1204
Avenue de l'Université, 64000 Pau, FRANCE

ABSTRACT

We report the presence of oscillations in cooperative graded-response neural networks which are devoid of negative connections. Our analysis of the asymptotic behavior of these networks shows that i) trajectories of most initial conditions tend to stable equilibria, ii) undamped oscillations are unstable, and can only exist in a narrow region forming the boundary between the basins of attraction of the stable equilibria. This analysis corroborates the hypothesis that the oscillations are transients. In fact, it is shown that the transient behavior of the system with delay follows that of the corresponding discrete-time network. The latter may display infinitely many non-constant periodic oscillations that transiently attract the trajectories of the network with delay, leading to long-lasting transient oscillations. The duration of these oscillations increases exponentially with the inverse of the characteristic charge/discharge time of the neurons.

1. Introduction

Graded-response neural networks have been used as models of biological neural networks as well as in artificial neural network applications. These systems display a wide range of behavior. One topic of investigation has been to determine how convergent or oscillating dynamics arise, and clarify the dependence of these on network connectivity (Hirsch, 1989). One factor that has been shown to influence the dynamics of graded-response neural networks is the presence of interunit transmission times, referred to as delays (Marcus et al., 1991; Destexhe and Gaspard, 1993; Gilli, 1993; Finnochiaro and Perfetti, 1995). These reflect propagation time along axons, synaptic delay etc in biological systems or finite switching times and transmission times in hardware implementations of artificial neural networks. Delays have been shown to induce oscillations in networks that would otherwise be convergent. Some of these oscillations can be traced to well-known delay induced instabilities that occur in some dynamical systems, specially those with negative or mixed feedback (Bélair et al., 1996). Others arise through a different mechanism, referred to as delay-induced transient oscillations. In this report, we briefly review our results concerning the influence of delays on the transient behavior of graded-response neural networks.

2. Graded-Response Neural Networks

Each graded-response neuron is described by its activation at time t, denoted by $a_i(t)$, an output function $\sigma_i(a_i)$, a decay rate $1/\epsilon_i$ and a constant input K_i. W_{ij} and τ_{ij} represent, respectively, the connection weight and delay between neurons j and i. The behavior of an n-neuron network is governed by the following system of delay differential equations (DDEs):

$$\epsilon_i \frac{da_i}{dt}(t) = -a_i(t) + K_i + \sum_{j=1}^{n} W_{ij}\sigma_j(a_j(t - \tau_{ij})) \quad 1 \leq i \leq n. \tag{1}$$

Throughout this paper we assume that the two following hypotheses are satisfied.

Hypothesis 1. *The neuron transfer function σ_i is sigmoidal, i.e., it is a smooth strictly increasing function, bounded between two real numbers $m_i < M_i$, such that there is a unique point p_i such that $\sigma_i''(p_i) = 0$. Without loss of generality, we suppose $p_i = 0$ for all i.*

Examples of sigmoidal functions used in neural networks are $\tanh(\beta_i a_i)$, or $\arctan(\beta_i a_i)$.

Hypothesis 2. *The connection matrix $W = [W_{ij}]$ is positive ($W_{ij} \geq 0$ for all i,j) and irreducible i.e. it does not leave invariant any proper nontrivial subspace generated by a subset of the standard basis vectors for $I\!R^n$.*

Let $\tau_j = \max_{1 \leq i \leq n}\{\tau_{ij} : W_{ij} \neq 0\}$, then $S = \mathcal{C}[-\tau_1, 0] \times \cdots \times \mathcal{C}[-\tau_n, 0]$ is the phase space for DDE (1). For any initial condition ϕ in S, there exists a unique solution of DDE (1) defined for all $t \geq 0$.

3. Asymptotic Dynamics

In this section we briefly state some of the results in (Pakdaman et al., 1996) regarding the asymptotic behavior of irreducible excitatory networks. Detailed proofs can be found in the original report.

Under hypotheses 1 and 2, DDE (1) is cooperative irreducible with bounded trajectories. Therefore *i)* the stability of the locally stable equilibria is not affected by the presence of delay, *ii)* The system is almost convergent: most trajectories in the system with delay converge to the stable equilibria.

So that for irreducible excitatory networks, delays are "almost" harmless, in the sense that they do not affect the asymptotic dynamics of most trajectories. It can be shown that the set of solutions that do not converge to stable equilibria is in fact exactly the union of the boundaries of their basin of attraction. These solutions display oscillations that may be damped, when situated on the stable manifold of a saddle point located on a basin boundary, or undamped. In the particular case of ring neural networks, we have shown that undamped oscillations are necessarily asymptotically periodic (Pakdaman et al., 1997a, 1997b).

In summary, in irreducible excitatory networks, the delay has little effect on the asymptotic behavior of the system. However, the situation is different with respect to the transients.

4. Transient Behavior

It can be shown that the solutions of DDE (1) are transiently attracted by those of the corresponding difference equation:

$$a_i(t) = K_i + \sum_{j=1}^{n} W_{ij}\sigma_j(a_j(t - \tau_{ij})) \quad 1 \leq i \leq n. \tag{2}$$

In contrast with DDE (1), this difference equation displays infinitely many attracting undamped oscillations. Thus trajectories of (1) with initial conditions in the basin of one of the oscillatory solutions of (2) display transient oscillations.

Our numerical investigations show that the duration of such transients increases exponentially as $\epsilon_i \to 0$, which, formally, corresponds to increasing the delay. These numerical results have been confirmed by the analytical evaluation of the transient regime duration for a two-neuron network with piecewise constant output functions (Pakdaman et al., 1995).

5. Conclusion

We investigated the asymptotic and the transient dynamics of irreducible excitatory graded-response neural networks. We showed that the delay does not affect the asymptotic behavior of most trajectories in such systems, yet it leads to long-lasting transient oscillations whose duration increases exponentially with the delay, indicating that they can outlast observation windows in numerical investigations. Therefore, for practical applications, these transients cannot be distinguished from stationary oscillations.

It is argued that understanding the transient behavior of neural network models is an important complement to the analysis of their asymptotic behavior, since both living nervous systems and artificial neural networks may operate in changing environments where long-lasting transients are functionally undistinguishable from asymptotic regimes.

Acknowledgment

KP, CPM, CGR and JFV were partially supported by USP–COFECUB under project U/C 9/94. CPM was also partially supported by CNPq (the Brazilian Research Council). OA received support from the CNRS PNDR-GLOBEC program.

References

Bélair J., S.A. Campbell and P. van den Driessche (1996)
"Frustation, stability and delay-induced oscillations in a neural network model", *SIAM Journal on Applied Mathematics*, 56:245–255.

Destexhe A. and P. Gaspard (1993)
"Bursting oscillations from a homoclinic tangency in a time delay system", *Physics Letters A*, 173:386–391.

Finnochiaro M. and R. Perfetti (1995)
"Relation between template spectrum and stability of cellular neural networks with delay", *Electronics Letters*, 31:2024–2026.

Gilli M. (1993)
"Strange attractors in delayed cellular neural networks", *IEEE Transactions on Circuits and Systems_I: Fundamental Theory and Applications*, 40:849–853.

Hirsch M.W. (1989)
"Convergent activation dynamics in continuous time networks", *Neural Networks*, 2:331–349.

Marcus C.M., F.R. Waugh and R.M. Westervelt (1991)
"Nonlinear dynamics and stability of analog neural networks", *Physica D*, 51:234–247.

Pakdaman K., C. Grotta-Ragazzo, C.P. Malta and J.-F. Vibert (1995) "Delay-induced transient oscillations in a two-neuron network", Publicações Instituto de Física, Universidade de São Paulo, Brasil, *Technical report IFUSP/P-1181*.

Pakdaman K., C.P. Malta, C. Grotta-Ragazzo, O. Arino and J.-F. Vibert (1997a) "Transient oscillations in continuous-time excitatory ring neural networks with delay", *Physical Review E*, 55:3234–3248.

Pakdaman K., C.P. Malta, C. Grotta-Ragazzo and J.-F. Vibert (1997b) "Effect of delay on the boundary of the basin of attraction in a self-excited single graded-response neuron", *Neural Computation*, 9:319–336.

Pakdaman K., C.P. Malta, C. Grotta-Ragazzo and J.-F. Vibert (1996) "Asymptotic behavior of irreducible excitatory networks of graded-response neurons", Publicações Instituto de Física, Universidade de São Paulo, Brasil, *Technical report IFUSP/P-1200*.

POTENTIALITY OF COMPLEX BEHAVIOUR IN SINGLE NEURONS

LADISLAV ANDREY
*Institute of Computer Science
Academy of Sciences
Neural Network Section
Pod Vodárenskou věží
182 07 Prague 8
Czech Republic*
E-mail:andre@uivt.cas.cz.

ABSTRACT

It is shown analytically that the sigmoidal form of its transfer function is a sufficient condition for a single neuron to behave chaotically.

1. Introduction

The ubiquitous feature of the nervous system is the widespread occurence of complex dynamical behaviors. Examples range from the spike trains of single neurons to the fluctuating potentials of thousands of neurons measured from the surface of the scalp by EEG (Elbert et al., 1994; McKenna et al.,1992; Milton, 1996).

Chaotic behavior has been identified experimentally even in a single neuron (Aihara et al., 1987; Matsumoto et al., 1987; McKenna et al., 1992; Milton, 1996; Rapp et al., 1985). It has been shown that responses of model neurons as the simplified Caianello (1961), and Nagumo and Sato (1972) models with an influence of the refractoriness due to a past firing that decreases exponentially with time, can also be chaotic (Aihara et al., 1990). It has also been shown numerically with the Hodkin-Huxley equations (Hodgkin and Huxley, 1952) and in squid giant axons, that responses of a resting nerve membrane to periodic stimulation are not always periodic and can be understood as deterministic chaos (Aihara et al., 1987; Aihara et al., 1990; Matsumoto et al., 1987).

Recently neurodynamics based on a nonmonotone response function for single neurons has been described (Morita, 1993; Yoshizawa et al.,1993). Such neurons also can behave chaotically (Shuai et al., 1996; 1997).

The dynamics of single neurons plays a very important role in recent approaches to artificial neural networks. Model neurons with complicated dynamics are the composer elements of artificial neural networks. Such neural networks are called chaotic neural networks (Aihara et al., 1990; Bondarenko 1994).

So far there is only numerical evidence for chaotic behavior in biologically motivated neuron models, caused by a time delay (refractoriness) in the model (Aihara et al., 1990; Ichinose et al., 1992) or by the nonmonotonicity of transfer function of the model (Shuai et al., 1996; 1997).

In this paper we will show analytically that in the basic McCulloch-Pitts like neuron model (McCulloch and Pitts, 1943) with biologically derived sigmoidal response function (Eeckman and Freeman, 1986), the potentiality of chaotic dynamics emerges naturally.

2. Potentiality of Dynamic Chaos in a Single Neuron with the Sigmoidal Transfer Function

A standard neuron model is a simple threshold element transforming a weighted summation of the inputs into the output through a nonlinear transfer function with a threshold. In what follows we will use a generalization of classical McCulloch-Pitts neuron model (McCulloch and Pitts, 1943), in which instead of the unit step function the sigmoidal transfer function will be exploited. Then for the i-th neuron dynamics,

$$y_i = g\left(\xi_i\left(t_n\right)\right) = \frac{1}{1 + e^{-\lambda \xi_i}}, \qquad (1)$$

holds, where

$$\xi_i\left(t_n\right) = h_i - \Theta_i = \sum_j w_{ij} x_j(t_n) - \Theta_i. \qquad (2)$$

Here $y_i(t_{n+1})$ is the output of the i-th neuron at the discrete time t_{n+1}; g is the sigmoidal transfer function with a slope λ; Θ is the threshold of the i-th neuron; w_{ij} (for $i \neq j$) is the connection weight with which the firing of the j-th neuron affects the ith neuron and $h_i(t_n)$ is the local field of i-th neuron at the discrete time t_n. One can speak about the input dynamics (Eq. 2) and the output dynamics (Eq. 1).

Now a transition from the discrete time dynamics (Eq. 1) and (Eq. 2) to a continuous time dynamics can be easily made (Andrey, 1991). Denote $\Delta t = t_n - t_{n-1}$ and suppose $\Delta t \ll 1$. Then from (Eq. 2) we have $\frac{1}{\Delta t}\left(\xi_i\left(t_n\right) - \xi_i\left(t_{n-1}\right)\right) =$

$-\frac{1}{\Delta t}(\xi_i(t_{n-1}) - \xi_i(t_n))$ and in the limit, where $\Delta t \to 0$, we get for the input dynamics

$$\frac{d\xi_i}{dt} = -\alpha \left(\xi_i(t_{n-1}) - \frac{\partial E}{\partial y_i} \right), \tag{3}$$

where $\alpha = \frac{1}{\Delta t}$ is a parameter and E is the Hopfield-like energy function

$$E = -\frac{1}{2} \sum_{i,j} (w_{i,j} x_j y_i - \Theta_i y_i). \tag{4}$$

Analogously, from (Eq. 1) the time continuous output dynamics (Eq. 5),

$$\frac{dy_i}{dt} = -\alpha(y_i - g(\xi_i)). \tag{5}$$

Here we are especially interested in the output dynamics as we are looking for the potentiality of chaotic responses in such neurons. To this end we exploit properties of the sigmoidal transfer function. Namely, from (Eq. 1) one gets directly

$$\frac{dy_i}{d\xi_i} = \lambda y_i (1 - y_i), \tag{6}$$

which is a Ricatti-like equation.

To analyse the output dynamics of the i-th neuron take the Taylor expansion of the output variable y_i in time,

$$\begin{aligned} y_i(t_{n+1}) &= y_i(t_n) + \frac{dy_i}{d\xi_i} d\xi_i + \cdots = \\ &= y_i(t_n) + \frac{dy_i}{d\xi_i} \frac{d\xi_i}{dt} \Delta t + \cdots, \end{aligned} \tag{7}$$

where again we assume $\Delta t = t_{n+1} - t_n << 1$. After substistuting from (Eq. 3) and (Eq. 6) into (Eq. 7), neglecting terms with $(\Delta t)^k$, $k \geq 2$, one gets after some elementary calculations, suprisingly

$$y(t_{n+1}) = y(t_n) + 4ay(t_n)(1 - y(t_n)), \tag{8}$$

where

$$a = \frac{\lambda}{4} (\xi(t_{n+1}) - \xi(t_n)). \tag{9}$$

For the sake of simplicity we have omitted the index i, as the analysis is performed for the i-th neuron. For the same reason we can put $y(t_n) \equiv y_n$ and $\xi(t_n) \equiv \xi_n$. Then (Eq. 8) can be written in the form

$$y_{n+1} = y_n + 4ay_n(1 - y_n), \tag{10}$$

where

Introducing the substitution

$$a = \frac{\lambda}{4}(\xi_{n+1} - \xi_n). \tag{11}$$

$$z_n = 4ay_n - \frac{1+4a}{2}, \tag{12}$$

into (Eq. 10) one gets an equivalent form of logistic equation for z's, namely

$$z_{n+1} = c - z_n^2, \tag{13}$$

where $c = 4a^2 - \frac{1}{4}$ is the parameter and a is determined by (Eq. 11).

Note that the logistic map (Eq. 13), on the interval $(-1, 1)$, where c varies in the interval $[0, 2)$, coincides with the more accustomed logistic map (Eq. 14)

$$x_{n+1} = bx_n(1 - x_n), \tag{14}$$

mapping the interval $(1 - \frac{b}{4}, \frac{b}{4})$ into itself when $2 < b \leq 4$. Under the coordinate change $x = \left(\left(\frac{b}{4} - \frac{1}{2}\right) \times z + \frac{1}{2}\right)$ with the parameter identification $c = \frac{b^2}{4} - \frac{b}{2}$, both families (Eq. 13) and (Eq. 14) coincide (Collet and Eckmann, 1980).

It is well known that the quadratic maps (Eq. 13) and (Eq. 14) can exhibit chaotic behavior. It means there exists some critical value of parameter c_{crit} in (Eq. 13) or b_{crit} in (Eq. 14) such that for $c \geq c_{crit}$ or $b \geq b_{crit}$ solutions of (Eq. 13) or (Eq. 14) can be chaotic. For (Eq. 13) then one has $c_{crit} = 1.8284$, or for (Eq. 14) then $b_{crit} = 3.8284$ (Collet and Eckmann, 1980).

So far we have been able to show analytically that the output dynamics, or firing patterns, of the generalized McCulloch-Pitts like model of single neuron with the sigmoidal transfer function contains inherently deterministic chaos. In other words, chaotic behavior is natural in neuronal dynamics.

3. Consequences of Single Neuron Complex Dynamics

The output dynamics described in a first approximation by the logistic equation (Eq. 13) is much more complex may appear at the first sight. Namely the parameter c, from (Eq. 2) and (Eq. 9), is in general the function

$$c = c(\lambda, \Theta, n). \tag{15}$$

That means the firing pattern of the neuron depends crucially upon the basic characteristics of the sigmoidal transfer function, namely the threshold Θ and the slope λ. As already mentioned Θ represents the threshold (bias) and from the biological point of view is a measure of sensitivity. So if Θ increases then the sensitivity decreases and vice versa. On the other hand the parameter λ presents the steepness of sigmoidal transfer function and from the biological point of view this is connected with the excitability (Andrey, 1997). We will not go to details here but only mention case of onset of chaos in (Eq. 13). For this special case it follows immediately from (Eq. 2) and (Eq. 9)

$$\lambda_{crit} = \frac{4a_{crit}}{h_{crit} - \Theta_{crit}}, \qquad (16)$$

where $a_{crit} = \frac{1}{2}\sqrt{c_{crit} + \frac{1}{4}}$ and $c_{crit} = 1.8284$ (Collet and Eckmann, 1980). Surprisingly enough at the onset of potential chaotic behavior we have a functional dependence between basic characteristics of neuronal dynamics, namely $\lambda = \lambda(\Theta)$ (Andrey, 1997a, b).

4. Conclusions

We have demonstrated, analytically, for the first time the potentiality of chaotic behavior in the single McCulloch-Pitts like neuron model with the sigmoidal transfer function. Such neurons can generate cyclic firing patterns of all possible periods, bursting patterns, as well as aperiodic patterns, and all possible combinations of these.

Acknowledgment

This research was partly supported by the Grant Agency of Czech Republic Grant GA-201/950992.

References

Aihara, K. and G. Matsumoto (1987). *Chaos in Biological Systems*, H. Degn, A.V. Holden and L. F. Olsen, eds., New York, Plenum, p. 121-131.
Aihara, K., T. Takabe and M. Toyoda (1990). *"Chaotic neural networks"*, Phys. Lett. A 144: 333-340.
Andrey, L. (1986). *" The relationship between entropy production and K-entropy"*, Prog. Theor. Phys. 75: 1258-60.

Andrey, L. (1991). *" Simple biological neural networks and stability criteria for equilibrium memories"*, Neurocomputing 3: 221-230.

Andrey, L. (1997). *"λ − Θ neural networks"*, NOLTA'97, it will be published.

Andrey, L. (1997). *Biological findings supporting a functional dependence λ upon Θ as well as a rigorous theoretical derivation* - see Techn. Rep. V-711, ICS, Prague.

Bondarenko, V.E. (1994). *" A simple neural network model produces chaos similar to the human EEG"*, Phys. Lett. A 196: 195-200.

Caianello, E.R. (1961). *"Outline of a theory of thought-processes and thinking machines"*, J. Theor. Biol. 2: 204-35.

Collet, P. and J.P. Eckmann (1980). *Iterated Maps on the Interval as Dynamical Systems*, Basel: Birkhäuser.

Eeckman, F.H. and W.J. Freeman (1991). *"Asymmetric sigmoid nonlinearity in the rat olfactory system"*, Brain Research 557: 13-21.

Elbert, T., et al. (1994). *"Chaos and physiology: Deterministic chaos in excitable cell assemblies"*, Physiol. Rev. 74: 1-40.

Hodgkin, A.L. and A.F. Huxley (1952). *" A quantitative description of membrane current and its application to conduction and excitation in nerve"*, J. Physiol. 117: 500-44.

Ichinose, N., K. Aihara and K. Judd (1992). *"Chaotic dynamics od a simple excitable neuron model"*, in: International Joint Conference on Neural Networks, Beijing, p. I-622-25.

Matsumoto, G., et al. (1987). *"Chaos and phase-locking in normal squid axon"*, Phys. Lett. 123: 162-6.

Mc Culloch, W.S. and W.H. Pitts (1943). *"A logical calculus of the ideas immanent in nervous activity"*, Bull. Math. Biophys. 5: 115-33.

Mc-Kenna, T., J. Davis and S. F. Zornetzer, ed. (1992). *Single Neuron Computation*, Boston: Academic Press.

Milton, J. (1996). *Dynamics of Small Neural Populations*, Providence: AMS.

Morita, M. (1993). *"Associative memory with nonmonotone dynamics"*, Neural Networks 6: 115-26.

Nagumo, J. and S. Sato (1972). *"On a response characteristic of a mathematical neuron model"*, Kybernetik 10: 155-64.

Rapp, P.E., et al. (1985). *"Dynamics of spontaneous neural activity in the simian motor cortex: the dimension of chaotic neurons"*, Phys. Lett. 6: 335-8.

Shuai, J.W., et al. (1996). *" A chaotic neural network model"*, Chin. Phys. Lett. 13: 185-7.

Shuai, J.W., et al. (1997). *"Chaos and hyperchaos in the nonmonotonic neuronal model"*, Neural Networks in press.

Yoshizawa, S., M. Morita and S. I. Amari (1993). *"Capacity of associative memory using a nonmonotonic neural model"*, Neural Networks 6: 167-76.

DETERMINISM VERSUS STOCHASTICITY IN BIOLOGICAL INTERSPIKE INTERVAL TIME SERIES

NICO STOLLENWERK

ForschungsZentrum Jülich, MOD, HLRZ,
D-52425 Jülich, Germany
E-mail: N.Stollenwerk@fz-juelich.de

ABSTRACT

Time-series of interspike intervals from intracellular recordings in spontaneously firing hippocampal CA1 cells in slice preparations from rabbits are investigated. Previously reported short time predictability can be explained purely by bursts occurring in otherwise random event distributions. Then from spontaneous activity conclusions are drawn for modelling approaches of hippocampal dynamics. Effective couplings between hippocampal compartments are found to be as well positive as negative. In such situations of not only positive couplings chaotic dynamics can arise with further implications for functional models.

1. ISI-Data Analysis

In recent years the question of determinism or pure stochasticity in interspike interval (ISI) time series from hippocampal slice preparations has been discussed intensively (Menéndez de la Prida et al., 1997). Here time-series of interspike intervals from intracellular recordings in spontaneously firing hippocampal CA1 cells from rabbits are investigated (see Fig. 1 a) for locating the hippocampal regions, after Ramón y Cajal[1]). The experimental conditions are described recently (Menéndez de la Prida et al., 1996). Slice preparations in three different experimental conditions are tested: *immature* from young rabbits up to seven postnatal days, *mature* and *bicuculline*-treated slices from adult rabbits. Bicuculline is a convulsion inducer through inhibition blockade. For a detailed description of the statistical analyses see a complete paper (Stollenwerk et al., 1997); the basic results are as follows:

Preparations from immature rabbits show the dynamical behavior of a Poisson process, while those from mature animals are periodically firing with additive Gaussian noise. In Fig. 1 b a return plot of interspike intervals is shown,

---

[1]Santiago Ramón y Cajal (1904) *Textura del sistema nervioso del hombre y de los vertebrados*, Tomo II, Imprenta Nicolás Moya, Madrid.

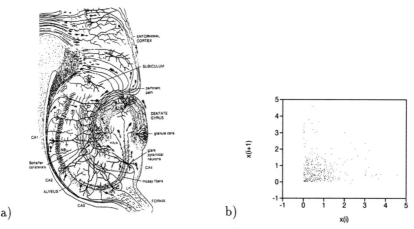

Figure 1: In a) the picture of a slice with some cells stained shows the basic histological structure of the hippocampus. In b) a typical return map of interspike intervals measured intracellularly in CA1 pyramidal neurons is shown.

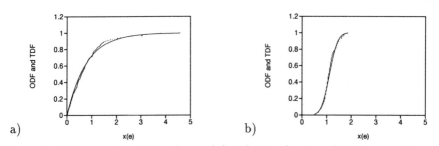

Figure 2: a) Comparison of the observed distribution function ODF from a data set from immature rabbits (marked by points) with the theoretical distribution function TDF (full lines) obtained from our null hypotesis of a Poisson process. b) The ODF for mature rabbits is much better described by the S-shaped TDF of a Gaussian process as underlying hypothesis than with a Poisson TDF, as was used in a).

typical for immature rabbits. In Fig. 2 estimates of the observed distribution functions (ODFs) are compared with the theoretical distribution functions (TDFs) for the simplest stochastic processes, i.e. in the case of immature the Poissonian distribution (Fig. 2 a) and for mature the Gaussian (Fig. 2 b). These simple processes cannot be rejected for the present data sets. Bicuculline-treated, hence inhibition blocked slices from mature animals, show again the same dynamics as from immature, namely Poissonian noise. So during maturation the inhibitory system develops to a dynamically important component in the whole hippocampal network.

In some immature and bicuculline preparations large proportions of bursts occur in between Poissonian distributed single spikes. By correcting these bursts away (Menéndez de la Prida *et al.*, 1997) we find better agreement with the null hypothesis of Poisson processes. See the distribution functions in Fig. 3 a for the original data and Fig. 3 b for the burst corrected data.

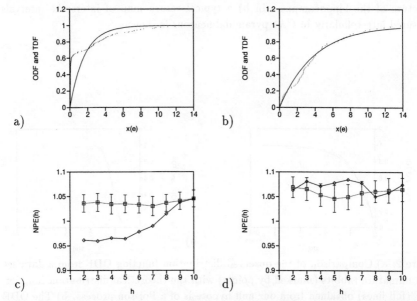

Figure 3: A data set deviating largely from simple null hypotheses (see a) and with signs of short time predictability (see c) is shown here. However, a simple burst correction algorithm clarifies the nature of the distribution to be Poissonian (see b) and removes the short time predictability (in d).

In some cases the frequent bursts lead to signs of short time predictability comparable with results stated by other authors. Short time predictability is assumed to characterize deterministic chaos. Here such findings are explained by quite trivial reasons: the different time scales of ISI in bursting and in spiking parts of the time series. Fig. 3 c shows the results of the predictability test for the original data. However, we can also remove this predictability by burst correction. Fig. 3 d presents the predictability result for the burst corrected data.

To test for stochasticity we use the Kolmogorov-Smirnov test which is especially well suited for such small data sets like those in the present experiments (Stollenwerk et al., 1997). Direct graphical analysis of the distribution function provides useful qualitative comparison. The predictability tests for nonlinearity are applied to the original data and its Gauss-scaled surrogates to test for the null hypothesis of linearly correlated process (see the squares with error bars in Fig. 3 c and d). The maximal distance between prediction error of the original data and of its surrogates, called significance or S-value[2], for all experimental data sets is given elsewhere (Stollenwerk, 1997), to confirm quantitatively our statements. Using 40 surrogate sets the value of the prediction test in Fig. 3 c is $S = 4.31$ while for Fig. 3 d it is $S = 1.36$ at an optimal burst correction of $\delta = 0.17$ sec. (see Menéndez de la Prida et al., 1997, for the definition of the burst correction parameter δ).

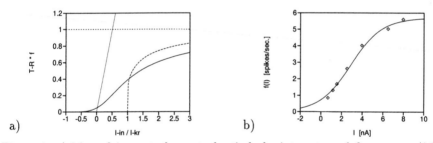

Figure 4: a) Mean firing rate for a stochastic leaky integrate and fire neuron (full line) as function of input current. The curved dashed line shows the same for a deterministic neuron, the dotted line without leakage (linear neuron). b) The sigmoidal shape of the mean firing rate under noise can be replaced by the standard sigmoidal in investigations of the dynamics (topologically equivalent). It even fits quite well experimental data (diamonds).

That the hippocampus is capable of showing not only noisy fixed point and periodic dynamics, has been demonstrated recently on other levels than single spike activity. Periodic forcing and measurements on the mass activity of many neurons have been used in such experiments both in slice preparations

---

[2]S is the distance between prediction error of the original data and mean error of the surrogate ensemble in units of the standard deviation of the surrogate prediction errors.

using extracellular electrodes (Hayashi & Ishizuka, 1995), and in experiments with awake rats performed for epilepsy studies, electrodes implanted for several weeks (Suzuki & Stollenwerk, 1997, and Stollenwerk, 1997).

2. Implications for Modelling Hippocampal Activity

Conclusions about the dynamical interplay between excitatory and inhibitory networks can be drawn from spontaneous activity as investigated in these experiments, since the reciprocal of the mean ISI estimates the mean instantaneous firing rate depending on the inputs to the neurons under investigation.

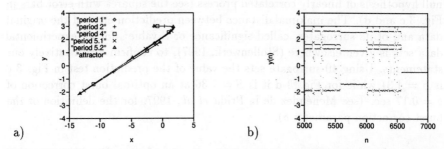

Figure 5: a) Chaotic attractor of a sigmoidal network with excitatory and inhibitory connections. On the attractor there are also shown the first periodic orbits, which can be stabilized with a controlling network of the same type of neurons in one consistent neural network. b) Time series of attractor switching by noise in such a consistent self-controlling network.

Some well known theoretical facts should be remembered here: A sigmoidal function from a noisy leaky integrate and fire neuron (analytically solvable only as a first passage time problem for Orstein-Uhlenbeck processes[3]) is obtained as mean firing rate as function of input current, see Fig. 4 a. Topologically sufficient for qualitative dynamical analyses is the standard sigmoidal function $\sigma(x) = 1/(1+e^{-x})$ (e.g. stability and Lyapunov exponents are preserved under transformations from any sigmoidal to $\sigma(x)$). The standard sigmoid even fits quite well experimental biological data[4]. From this fit (Fig. 4 b) as well as from the theoretical curve it can be seen that for zero input current the mean firing rate is nonvanishing.

The step from single neuron models to compartments with weighted stochastic connections preserves dynamical properties as e.g. whole bifurcation dia-

---

[3]Ricciardi, L.M., & S. Sato (1988) First-Passage-Time Density and Moments of the Ornstein-Uhlenbeck Process, J. Appl. Prob. 25, 43–57. See also Stollenwerk, 1997.

[4]Data from Connor, J.A., & C.F. Stevens (1971) "Prediction of repetitive firing behaviour from voltage clamp data on an isolated neurone soma", J. Physiol. 213, 31–53.

grams (Wennekers & Pasemann, 1997). Now, a parametrization of σ gives for spontaneous activity in the different hippocampal compartments (the dentate gyrus, CA3 and CA1, see Fig. 1 a) effective connections with positive and negative sign (recent experiments from Menéndez de la Prida & Sánchez-Andrés, 1997) namely effective self-inhibition in CA1. Under such conditions, already two sigmoidal neurons (respectively compartments) can exhibit chaotic dynamics (Pasemann & Nelle, 1993, Pasemann, 1995).

This leads us to theoretical considerations of chaotic attractors in combined excitatory and inhibitory networks and to a novel mechanism of multiple attractor neural networks on the basis of chaos control algorithms (Fig. 5), in which the interplay of nonlinear determinism and stochasticity plays a key role (Stollenwerk & Pasemann, 1996 (a),(b)). Attractor switching between periodic orbits is achieved through a transient along a chaotic attractor. As well the chaotic module as the controller consist of the same type of sigmoidal neurons in a consistent dynamics. In the future such a mechanism of attractor switching might help to understand dynamic memories, as they are assumed to be present as a mayor function in the hippocampus.

Acknowledgments

For providing the data and many interesting discussions I acknowledge Liset Menéndez de la Prida and Juan Vicente Sanchez-Andrés, Alicante, Spain. I would also like to thank Friedhelm Drepper, Jülich, Germany, and Frank Pasemann, Leipzig, Germany, for many technically helpful suggestions, as well as John Taylor, London, UK, for drawing my attention to links between physiological and functional modelling.

References

Hayashi, H., and S. Ishizuka (1995) "Chaotic responses of the hippocampal CA3 region to a mossy fiber stimulation in vitro", *Brain Research*, **686**, 194–206.

Menéndez de la Prida, L., S. Bolea and J.V. Sánchez-Andrés (1996) "Analytical characterization of spontaneaous activity evolution during hippocampal development in the rabbit", *Neurosci. Lett.* **218**, 1–3.

Menéndez de la Prida, L., and J.V. Sánchez-Andrés (1997). "Mechanisms of Synchronization in the Hippocampus and its Role along Development", *contribution to IWANN '97, Lanzarote*, (pp. 154–161), Lecture Notes in Computer Sciences 1240, eds. J. Mira, R. Moreno-Díaz, J. Cabestany.

Menéndez de la Prida, L., N. Stollenwerk and J.V. Sánchez-Andrés (1997) "Bursting as a source for predictability in biological neural network activity", *Physica* **D 110**, 323–331.

Pasemann, F. (1995) "Neuromodules: A dynamical systems approach to brain modelling", in: *Supercomputing in Brain Research - From Tomography to Neural Networks* eds. H. Herrmann, E. Pöppel, D. Wolf, (World Scientific, Singapore), 331–347.

Pasemann, F. and E. Nelle (1993) "Elements of nonconvergent neurodynamics," in *Dynamical Systems, Theorie and Applications,* eds. S.I. Andersson, A.E. Andersson, U. Ottoson, (World Scientific, Singapore) pp. 167–201.

Stollenwerk, N. (1997) *Nichtlineare Dynamik in der Analyse neurowissenschaftlicher Systeme,* Dissertation, TU Clausthal.

Stollenwerk, N., L. Menéndez de la Prida and J.V. Sánchez-Andrés (1997) "Spontaneous Activity of Hippocampal Cells in Various Physiological States", *contribution to IWANN '97, Lanzarote, June, 3–10,* (pp. 91–102), Lecture Notes in Computer Sciences 1240, eds. J. Mira, R. Moreno-Díaz, J. Cabestany.

Stollenwerk, N., and F. Pasemann (1996(a)) "Control Strategies for Chaotic Neuromodules", *International Journal of Bifurcation and Chaos,* **6**, 693–703.

Stollenwerk, N., and F. Pasemann (1996(b)) "Switching in Self-Controlled Chaotic Neuromodules", *World Congress on Neural Networks, San Diego, California,* Sept. 15-18, 1996, (pp. 680–684). INNS Press and Lawrence Erlbaum, New Jersey.

Suzuki, T., and N. Stollenwerk (1997) "Chaotic response to periodic stimulation in hippocampus in freely moving rats relates to epileptic behavior", *in preparation.*

Wennekers, T., & Pasemann, F. (1996) "Synchronous Chaos in Highdimensional Modular Neural Networks", *International Journal of Bifurcation and Chaos* **6**, 2055–2067.

SYNCHRONIZATION AND CHAOS IN NETWORKS CONSISTING OF INTERACTING BISTABLE ELEMENTS

VLADIMIR CHINAROV

Sci.Res.Cntr Vidhuk, Vladimirskaya Str. 61-b, 252033 Kiev, Ukraine
E-mail: chinarov@vidguk.freenet.kiev.ua

ABSTRACT

Phase synchronization of activity patterns in ensembles consisting of interacting bistable elements, and chaotic dynamics of the network driven by periodic input were studied. It was shown that oscillatory dynamics of a system of phase oscillators with cosine interaction between them may give different patterns of synchronization of activities within clusters as well as clustering of attractors. Different types of the network architecture such as mean-field and nearest-neighbor interaction, symmetric and asymmetric types of coupling among elements were used.

1. Introduction

The problem of synchronization of coupled non-linear oscillators have attracted recently much attention (Tass & Haken, 1996; Fukai, 1996; Arnoldy & Brauer, 1996; Steriade *et al.*, 1995). There has been a great deal of recent work on the simulation of synchronization in neural networks consisting of coupled bistable elements that appear both in spatially homogeneous and heterogeneous structures with symmetric and asymmetric connections among elements, and may arise due to the spatial and temporal character of non-linear interaction between them. The attention in this field is focused mainly on the dynamical behavior of neural networks consisting of excitatory and inhibitory neuron populations, dynamic pattern formation related to neural oscillation, stochastic modeling of memory search, and the like (Chinarov, 1995; Chinarov *et al.*, 1996). The purposes of this paper is to investigate the dynamic pattern formation related to ensemble oscillations and to present the results of modeling non-equilibrium dynamics in a network consisting of interacting non-linear phase oscillators. Pattern formation and cooperative dynamics of periodically driven system is also studied.

2. The Model

A model that describes the non-equilibrium network dynamics for interacting Josephson-like elements with lateral excitatory and inhibitory couplings was studied analytically and by computer simulation. The dynamics of such a system is described by the equations for the phase oscillators with cosine interaction between them (Chinarov et al., 1996). Two cases are considered here: linear coupling among elements and discrete type of linear diffusion coupling (nearest-neighbor coupling) for the phases of underdamped and overdamped oscillators. For this model an extremely rich and unusual dynamic behavior including strange attractors, long-term and short-term metastable states was observed.

This model may be considered as a generalization of the model of attention developed earlier (Kazanovich et al., 1991) that characterized by central oscillators (CO) connected with peripheral oscillators (PO). There are no connections between PO's. We have used here different types of the network architecture, namely, mean-field coupling, nearest-neighbor coupling, symmetric (asymmetric) connection among elements. This network may be also forced by periodic input signal. The dynamics of this kind of network is described by the following equation

$$\ddot{q}_i(t) = -f_i - \eta \dot{q}_i(t) + \sin(\varphi_i(t) - q_i(t)) + \frac{a_i^q}{N} \sum_j \sin(q_i - q_j) e^{-\alpha|i-j|} + A\cos(2\pi f t),$$

$$\dot{\varphi}_i(t) = \varphi_i^0 - a_i^\varphi \sin(\varphi_i(t) - q_i(t)), \qquad (1)$$

where q_i and φ_i are the phases of coupled oscillators, η is a friction coefficient, α is a site-correlation length, A and f are the amplitude and frequency of the periodic input signal, ε is a coupling strength. The mean-field coupling among PO elements is described by the following equation

$$\dot{\varphi}_i(t) = \varphi_i^0 - a_i^\varphi \sin(\varphi_i(t) - <q_i(t)>), \qquad (2)$$

where brackets define an ensemble averaging.

3. Results of Synchronization Pattern's Simulation

Studying the dynamics of a system consisting of phase oscillators with cosine interaction between them, we may observe synchronization patterns of activities in elements within clusters remote from one another as well as clustering of different attractors. Examples of synchronization patterns that appear in different clusters of oscillators are shown in Fig. 1 on the phase plane for different values of network parameters. The mean field here is constructed by summarizing activities of CO's and PO's. There might exist different clusters having opposite phases of activity.

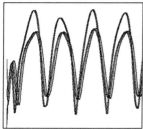 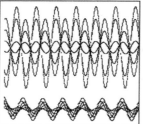 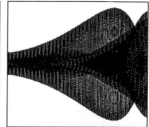

Fig. 1 Synchronization in clusters via two types of the mean field coupling – short-range (among q oscillators) and long-range (between q and φ nets). Left panel: 3- and 5-cluster attractors. Period-3 oscillations ($A=1$, $f=0.5$). Middle panel: Anti-phase synchronization between 2 cluster nets ($A=0$, Top: q-oscillators, Bottom: φ-oscillators). Right panel: Example of oscillatory pattern ($A=0$, $\eta=0.01$, $a^q=10.55$, $a^\varphi=1$ $\alpha=0$).

Cluster synchronization of periodically driven system as well as resonance-like synchronization of activities of the network where only one of the q oscillators is driven may be observed too.

When the mean-field interaction is switched on (see Eq. 2), one may observe even the synchronization of random bistable elements that model the neuron dynamics. Such a case is shown in Fig. 2 for three coupled bistable elements that switch in a random manner between two stable states. It is clearly seen that for weak values of coupling strength (left panel) there is no synchronization, while for the stronger one (right panel), there is a strong synchronization of random dynamics. It may be shown also that the coupling strength may be considered in this case as an order parameter.

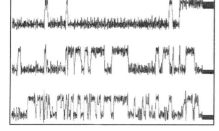 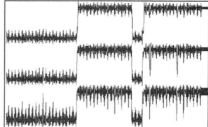

Fig. 2 Synchronization of random bistable elements. Left panel – coupling parameter = 0.075, right panel – coupling parameter = 0.75.

The strange attractors, example of which, as shown in Fig. 3, may also be found when all elements are synchronized within two clusters for the case when there are two mean fields – one for PO's and one for CO's. All the trajectories are localized finally at some region of the phase plane having stochastic nature.

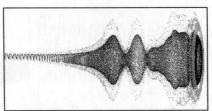

Fig. 3 Strange attractor on the phase plane for PO oscillators. The oscillators are synchronized within two clusters (that are seen as patterns with different density). These clusters are anti-synchronized.

4. Conclusion

Different patterns of synchronization were found for the system of coupled phase oscillators with cosine interaction between elements. It was shown that synchronization phenomena occur in system with short and long range mean field connections (Fig. 1). The synchronization of coupled random bistable elements has also been observed (Fig. 2). Clustering of attractors with chaotic dynamics (Fig. 3) may be found in the type of a system considered here.

References

Arnoldy, H.-M. R. and W. Brauer. (1996). "Synchronization without oscillatory neurons", *Biol. Cybern.* 74: 209–223.

Chinarov, V. (1995). "Stochastic modelling of memory search in neural networks", in: *System Analysis, Modelling, Simulation*, A. Sydow, ed., Amsterdam: Gordon & Breach Publ., Vol. 18–19, 301–304.

Chinarov, V., T. Gergely, and Yu. Kirpach. (1996). "Modelling of nonequilibrium dynamics in interacted ensembles of neural nets", in: *Proc. of CESA'96 Intern. Conf., Lille, France, 9–12 July*, P.Borne, G. Dauphin-Tanguy, C. Sueur, and S. El Khattab., eds. Vol. 2, 1250–1255.

Fukai, T. (1996). "Bulbocortical interplay in olfactory information processing via synchronous oscillations", *Biol.Cybern.* 74:309–317.

Kazanovich, Ya. B., V. I. Kryukov, and T. V. Lyuzyanina. (1991). "Synchronization and phase locking in oscillatory models of neutral networks", in: *Neurocomputers and Attention*, A. V. Holden and V. I. Kryukov, eds, Manchester: University Press, Vol. 1, 269–284.

Steriade, M., F. Amzica, and D. Contreras. (1996). "Synchronization of fast (30–40 Hz) spontaneous cortical rhythms during brain activation", *J. Neurosci.* 16: 392–416.

Tass, P. and H. Haken. (1996). "Synchronized oscillations in the visual cortex – a synergetic model", *Biol. Cybern.* 31–39.

STOCHASTIC MODEL OF POPULATION DYNAMICS

FERDINANDO DE PASQUALE
Dipartimento di Fisica, Università di Roma, La Sapienza, P.le A. Moro 2, I-00185 Roma, Italia
E-mail: depasquale@roma1.infn.it

and

BERNARDO SPAGNOLO
INFM, Unità di Palermo, and Dipartimento di Energetica ed Applicazioni di Fisica, Università di Palermo, Viale delle Scienze, I-90128, Palermo, Italia
E-mail: spagnolo@unipa.it

ABSTRACT

We study the population dynamics of an ecosystem by means of a simple stochastic model: a N-species generalization of the Lotka-Volterra model with a Malthus-Verhulst type of self regulation mechanism. The influence of the environment is taken into account by using a multiplicative noise and two types of interaction between the species are considered: a) mean field approximation b) random. The role of the noise on the transient dynamics and on the stability-instability transition is analyzed. We obtain asymptotic behaviour of the time average of the ith species due to the random field of the other species for three different regions of the growth parameter δ.

1. Introduction

Recently much work has been devoted to apply models from statistical physics to problems of evolution in biological systems (Higgs, 1995; Kauffman, 1993; Mezard et al., 1987). The present paper reports the study of a simple model of a large ecological system in which each species consists of a population interacting with the others in the framework of mean field theory and with random interaction. We consider an N-species generalization of the usual Lotka-Volterra model with a Malthus-Verhulst modelization of the self regulation processes for a fully connected ecological network. In this model the species extinction is not prevented as in the Gompertz model (Rieger, 1989). The generalized Lotka-Volterra equations model interactions between biological species and chemical reactions, and are also employed in laser physics, hydrodynamics and neural networks (Morita, 1982; Cairó & Feix, 1992). In a very recent paper the Lotka-Volterra equations are used to model the evolution and the extinction of a biosystem in a very long time scale (Abramson, 1997). The main motivation here arises from the study of complex ecosystem models such as natural ecosystems (Ayala, 1969) and idiotopic network in the immune system (Farmer et al., 1986) for which an equilibrium population exists, starting from a given initial distribution of the species. In our study we use a multiplicative noise to take into account the influence of the environment (climate, disease, etc..) and we investigate the nonlinear relaxation for two different types of interaction between the species.

(a) The species interaction is introduced by using a mean field approximation, i. e. the growth parameter is proportional to the average species concentration (Spagnolo, 1995; Ciuchi et al., 1996). In the large N limit, with N being the species number, the average of the species concentration has negligible fluctuations and the stochastic differential equation can be solved exactly.

The role of the noise on the nonlinear relaxation and on the stability-instability transition is analyzed. Our theoretical results reproduce not only the asymptotic behaviour but almost all the transient evolution of the system.

(b) The interaction between the species is assumed to be random. We study the asymptotic regime for the three different values of the parameter δ, which describes the development of the ith species without interacting with other species. We use again an approximation for the time integral of the average species concentration (Ciuchi et al., 1996), which greatly simplifies the stochastic evolution of the system.

The paper is organized as follows. The model is described in the next section. In section 3 we give some exact results for the two types of interaction and for the case (a) we compare the teoretical results with numerical simulations. In section 4 we draw the conclusions. We mainly focus our investigation on the role of the noise on the transient dynamics.

2. The model

The Malthus-Verhulst (MV) model was originally introduced to take into account a self regulation mechanism which prevents exponential growth of a simple population in absence of interaction with other species. To allow for species interaction we consider a generalization of the Lotka-Volterra model for a fully connected ecological network of N interacting species:

$$dn_i(t) = n_i(t) \left[\left(F(n_i(t)) + \sum_{j=1}^{N} J_{ij} n_j(t) \right) dt + \sqrt{\epsilon} dw_i \right] \quad (1)$$

where the function

$$F(n_i) = \frac{\delta + \frac{\epsilon}{2} - \gamma n_i^\alpha}{\alpha} \quad (2)$$

describes the development of the ith species without interacting with other species. In Eq.(1) the parameters γ and δ identify the saturation effects and the growth of the population; the interaction matrix J_{ij} modelizes the interaction between different species ($i \neq j$) and w_i is the Wiener process whose increment dw_i satisfies the usual statistical properties

$$< dw_i(t) > = 0; \quad < dw_i(t) dw_j(t') > = \delta_{ij} \delta(t-t') dt. \quad (3)$$

For $\alpha = 0$ we have the Gompertz model, while for $\alpha = 1$ we have the MV model, which we analyze here. In Eq. (1) we adopt the Ito prescriptions. The solution of Eq.(1) is given by

$$n_i(t) = \frac{n_i(0) \exp\left[\delta t + \sqrt{\epsilon} w_i(t) + \int_0^t dt' \sum_{j \neq i} J_{ij} n_j(t')\right]}{1 + \gamma n_i(0) \int_0^t dt' \exp\left[\delta t' + \sqrt{\epsilon} w_i(t') + \int_0^{t'} dt'' \sum_{j \neq i} J_{ij} n_j(t'')\right]}. \quad (4)$$

3. Mean field theory and random interactions

(a) We first consider symbiotic interaction between the species ($J > 0$) and mean field approximation. Our stochastic differential equation (SDE) becomes

$$dn_i = \left[\left(Jm + \delta + \frac{\epsilon}{2} \right) n_i - \gamma n_i^2 \right] dt + \sqrt{\epsilon} n_i dw_i \quad (5)$$

where

$$\sum_{j \neq i} J_{ij} n_j(t) = \frac{J}{N} \sum_j n_j(t) = Jm(t) \quad (6)$$

and m(t) is the site average. In the limit of a large number of interacting species the stochastic evolution of the system is given by the following integral equation

$$M(t) = \frac{1}{N\gamma} \sum_i \ln\left(1 + \gamma n_i(0) \int_0^t dt' e^{JM(t')+\delta t'+\sqrt{\epsilon}w_i(t')}\right), \qquad (7)$$

where

$$M(t) = \frac{1}{N} \sum_i \int_0^t dt' n_i(t') = \int_0^t dt' m(t'). \qquad (8)$$

is the time integral of the site population concentration average. We introduce an approximation of this Eq.(7) which greatly simplifies the noise affected evolution of the system and allows us to obtain analytical results for the population dynamics. We note that in this approximation the noise influence is taken into account in a nonperturbative way, and that the statistical properties of the time average process M(t) are determined asymptotically from the statistical properties of the process $w_{max}(t) = \sup_{0<t'<t} w(t')$, where w is the Wiener process (Ciuchi et al., 1993; Ciuchi et al., 1996). In Fig. 1 the analytical results are compared with the numerical solution of the Langevin equation (5), using n = 1000 species. In this figure we report the transient behaviour of the first moment of the time average $M(t)/t$ when the system relaxes towards the absorbing barrier ($\delta = -0.01 < 0$), with the following parameter setting: $J = 1$, $\gamma = 3$, $\epsilon = 0.01$, $m(0) = 1$, $\sigma^2_{n_i(0)} = 0.01$. The solid lines are the results of the theory. The dotted lines are the numerical results.

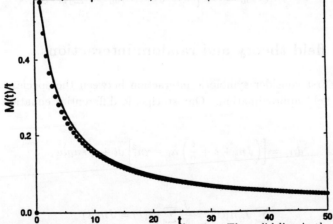

Fig. 1. Relaxation towards the absorbing barrier. The solid line is the result of the theory.

Starting from the following approximated integral equation for $M(t)$

$$M(t) \simeq \frac{1}{N\gamma}\sum_i \ln\left(1+\gamma n_i(0)e^{\sqrt{\epsilon}w_{max_i}}\int_0^t dt' e^{JM(t')+\delta t'}\right) \qquad (9)$$

it is possible to analyze the role of the noise on the stability-instability transition in three different regimes of the nonlinear relaxation of the system: (i) towards the equilibrium population ($\delta > 0$), (ii) towards the absorbing barrier ($\delta < 0$), (iii) at the critical point ($\delta = 0$). We obtain for the case (i) an explicit expression of the transition time t_c as a function of the noise intensity (ϵ), the initial population distribution ($n_i(0)$) and the parameters of the system (γ, J, δ)

$$t_c \simeq \frac{1}{\delta}\left\{\left[\frac{\epsilon}{2\pi\delta}+\ln\left(1+\left(\frac{\delta\gamma}{(J-\gamma)}\right)e^{-\langle\ln(\gamma n_i(0))\rangle}\right)\right]^{1/2}-\sqrt{\frac{\epsilon}{2\pi\delta}}\right\}^2 \qquad (10)$$

For the cases (ii) and (iii) we give two implicit expressions in terms of exponential and error functions. Particularly we have for the case (ii)

$$e^{-\left(\sqrt{|\delta|t_c}-\Delta_\epsilon\right)^2}+\sqrt{\pi}\Delta_\epsilon erf\left(\Delta_\epsilon-\sqrt{|\delta|t_c}\right)=\bar{\Gamma}(\epsilon,n_i(0)) \quad \text{for } \delta<0, \qquad (11)$$

where $\Delta_\epsilon = \sqrt{\epsilon/(2\pi|\delta|)}$, and

$$\bar{\Gamma}(\epsilon,n_i(0)) = e^{-\frac{\epsilon}{2\pi|\delta|}}\left(1-\frac{|\delta|\gamma}{(J-\gamma)e^{\langle\ln(\gamma n_i(0))\rangle}}\right)+\sqrt{\pi}\Delta_\epsilon erf(\Delta_\epsilon). \qquad (12)$$

For the case (iii) we obtain

$$e^{\sqrt{\frac{2\epsilon t_c}{\pi}}}\left(\sqrt{\frac{2\epsilon t_c}{\pi}}-1\right)=\frac{\gamma\epsilon}{\pi(J-\gamma)}e^{-\langle\ln(\gamma n_i(0))\rangle}-1 \quad \text{for } \delta=0. \qquad (13)$$

From Eq.s (10),(11), and (13) we obtain the behaviour of the time t_c as a function of the noise intensity (see Fig. 2). For a Gaussian peaked initial distribution of the populations, the transition time becomes independent of the $P(n_i(0))$ distribution.

From Fig. 2 we can see that the transition time increases from $\delta > 0$ to $\delta < 0$ according to

$$(t_c)_{\delta<0} > (t_c)_{\delta=0} > (t_c)_{\delta>0}. \qquad (14)$$

This means that when the interaction between the species prevails over

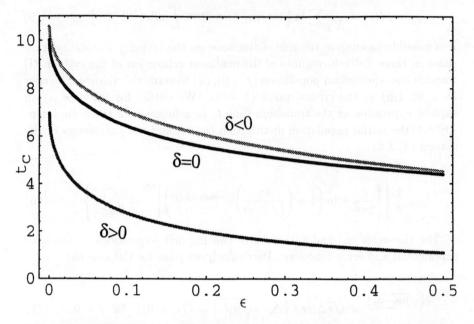

Fig. 2. Transition time t_c as a function of the noise intensity for three values of the growth parameter: $(a)\delta = -0.01, (b)\delta = 0, (c)\delta = 0.1$.

the resources, the presence of a hostile environment ($\delta < 0$) causes a late start of the divergence of some population (i.e. the instability).

The noise forces the system to sample more of the available range in the parameter space and therefore moves the system towards the instability. The effect of the noise is to make unstable the system earlier than in the deterministic case ($\epsilon = 0$ in Fig. 2).

The effect of a large variance of the Gaussian initial distribution of the populations (namely $\sigma^2_{n_i(0)} = 1$) on the transient behaviour of the time average $M(t)/t$ has been considered in a previous paper (Ciuchi et al., 1996).

The system recalls the initial distribution because of the term $\langle \ln(\gamma \varphi_i(0)) \rangle$ and because of the memory effect of the integral operator. As a consequence the process $M(t)/t$ is lowered for all the transient.

If we raise the intensity of the noise and keep fixed the initial distribution, we obtain the same effect of the enhancement of the variance of the Gaussian initial distribution of the population (see Fig. 3) for moderate values of noise intensity (namely $\epsilon = 0.1$).

For high values of noise intensity (namely $\epsilon = 1$) we strongly perturb the population dynamics and because of the presence of an absorbing barrier we obtain quickly the extinction of the populations (see Fig. 4).

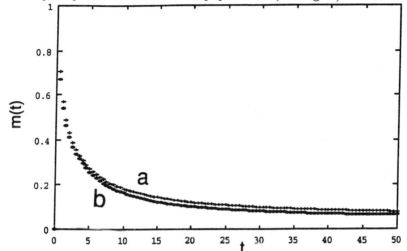

Fig. 3. Plot of $M(t)/t$ as a function of time for two values of the noise intensity: $(a)\epsilon = 0.01$ (cross); $(b)\epsilon = 0.1$ (dots), for $\delta = 0.1$ and $\gamma = 3$.

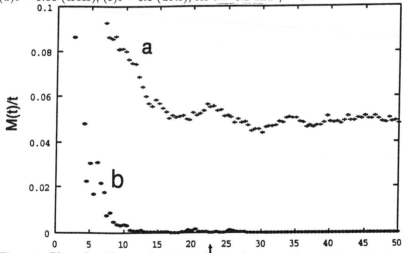

Fig. 4. Plot of $m(t)$ as a function of time for two values of the noise intensity: $(a)\epsilon = 0.01$ (cross); $(b)\epsilon = 1$ (dots), for $\delta = 0.1$ and $\gamma = 1.2$.

(b) We consider now random interactions between the species and focus on the statistical properties of the time integral of the ith population $N_i(t)$ in the asymptotic regime. From Eq.(1) we have

$$N_i(t) = \frac{1}{\gamma} \ln\left[1 + \gamma n_i(0) \int_0^t dt' exp\left[\delta t' + \sqrt{\epsilon} w_i(t') + \sum_{J \neq i} J_{ij} N_j(t')\right]\right], \quad (15)$$

where

$$N_i(t) = \int_0^t dt' n_i(t'). \quad (16)$$

In Eq. (15) the term $\sum_j J_{ij} N_j$ gives the influence of other species on the differential growth rate of the time integral of the ith population and represent a local field acting on the ith population. This term can be thought, following the Onsager's original idea of the cavity field (Mezard et al., 1987, pp.65–76), as the sum of two terms:

$$h_i = \sum_j J_{ij} N_j(t) = J\eta_i + \mu_i, \quad (17)$$

the cavity field ($J\eta_i$) and the reaction field (μ_i). The cavity field is the field acting on the ith population when this population is absent

$$J\eta_i = \sum_{j \neq i} J_{ij} N_j \quad (18)$$

In Eq.(18) N_j must be evaluated in absence of the ith population. Therefore the cavity field is a linear combination of the populations j$\neq$ i with random coefficients J_{ij}. So N_j does not depend on J_{ij} because such a dependence can be only due to the interaction with the ith population that is absent by definition. As a consequence the cavity field is a sum of many independent variables just because the ecological network is fully connected. By applying the central limit theorem it is easy to show that η_i is a Gaussian variable with zero mean and variance $\sigma_{\eta_i}^2 = J^2 \sigma_{N_i}^2$. Now we must calculate the effect of the presence of the ith population on the distribution of the jth populations. If such modification is small we can apply the standard linear response theory and we obtain

$$\tilde{N}_j = \int_0^t dt' \tilde{n}_j = \int_0^t dt' \int_0^{t'} dt'' G(t' - t'') J_{ji} n_i(t'') \quad (19)$$

where $\tilde{n}_j$ is the variation of n_j due to the presence of the ith population, and $G(t' - t'')$ is the response function. Thus the reaction field is given by

$$\mu_i = \sum_{j \neq i} J_{ij} \tilde{N}_j = \sum_{j \neq i} J_{ij} J_{ji} \int_0^t dt' \int_0^{t'} dt'' G(t' - t'') n_i(t''). \quad (20)$$

If we suppose that $G(t' - t'') = \Gamma\delta(t' - t'')$ and take into account that in the large N limit the term $\sum_j J_{ij}J_{ji}$ in Eq.(20) is a self averaging quantity, we get

$$\mu_i = \sum_{j \neq i} J_{ij}J_{ji}\Gamma N_i = J_o^2 \Gamma N_i(t) \qquad (21)$$

where J_o is the interaction intensity. From Eq.s (18), (20), and using the same approximation of the Eq.(9), after differentiating, we get the asymptotic solution of Eq.(15)

$$N_i(t) \simeq \frac{1}{\gamma - \Gamma J_o^2}\left[\ln\left[n_i(o)(\gamma - \Gamma J_o^2)\right] + \sqrt{\epsilon}w_{max_i}(t) + J\eta_i(t) + \ln\left(\frac{e^{\delta t} - 1}{\delta}\right)\right] \qquad (22)$$

which is valid for $\delta \geq 0$, that is when the system relaxes towards an equilibrium population and at the critical point. Evaluating Eq.(22) for $\delta = 0$, after making the ensemble average, we obtain for the time average of the ith population $\bar{N}_i$

$$\langle \bar{N}_i \rangle \simeq \frac{1}{(\gamma - \Gamma J_o^2)t}\left[N\sqrt{\epsilon t} + \ln t + \langle \ln\left[n_i(o)(\gamma - \Gamma J_o^2)\right]\rangle\right] \qquad (23)$$

which is a result consistent with that recently obtained using a mean field approximation (Ciuchi et al., 1996). When the system relaxes towards the absorbing barrier ($\delta < 0$) we get from Eq.s (15),(18) and (21), in the long time regime

$$N_i(t) \simeq \frac{1}{\gamma - \Gamma J_o^2}\left[\ln(\gamma n_i(0)) + \ln\left[\int_0^t dt' e^{\delta t' + \sqrt{t'}w_i(t') + j\eta_i(t')}\right]\right]. \qquad (24)$$

We see that the time integral of the population $N_i(t)$ is a functional of the cavity field and the Wiener process, in other words it depends on the history of these two stochastic processes.

4. Conclusions

We studied a stochastic model of an ecosystem of N interacting species. By means of an approximation of the integral equation, which gives the stochastic evolution of the system, we obtain analytical results very well reproducing almost all the transient. We investigate the role of the noise on the stability-instability transition and on the transient dynamics. For random interaction we obtain asymptotic behaviour for three different nonlinear relaxation regimes. A more detailed study of the random interaction case will be object of a future investigation.

Acknowledgements

This work is supported in part by the National Institute of Physics of Matter (INFM) and the Italian Ministry of Scientific Research and University (MURST).

References

Abramson, G. (1997) "Ecological model of extinctions", *Phys. Rev. E* 55: 785-788.
Ayala, F. J. (1969) "Experimental Invalidation of the Principle of Competitive Exclusion", *Nature* 224: 1076-1079.
Cairó, L. and M. R. Feix (1992) "Families of invariants of the motion for the Lotka-Volterra equations: the linear polynomial family", *Math. Phys.*33: 2440-2455.
Ciuchi, S., F. de Pasquale and B. Spagnolo (1993) "Nonlinear relaxation in the presence of an absorbing barrier", *Phys. Rev. E* 47: 3915-3926.
Ciuchi, S., F. de Pasquale and B. Spagnolo (1996) "Self Regulation Mechanism of an Ecosystem in a Non Gaussian Fluctuation Regime", *Phys.Rev.E* 53: 706-716.
Farmer, J.D., N. H. Packard and A. S. Perelson (1986) "The Immune System, Adaptation, and Machine Learning", *Physica D* 22:187-2204.
Higgs, P. G. (1995) "Frequency distributions in populations genetics parallel those in statistical physics", *Phys. Rev.* 51: 95-101.
Kauffman, S. A. (1993) *The Origin of Order: Self-Organization and Selection in Evolution*, Oxford University Press: London.
Mezard, M., G. Parisi and M. A. Virasoro (1987) *Spin Glasses Theory and Beyond*, World Scientific Lect. Notes in Physics 9: Singapore.
Morita, A. (1982) "An exact expression for the average value of the population governed by the stochastic Verhulst equation", *J. Chem.Phys.*76: 4191-4194.
Rieger, H. (1989) "Solvable model of a complex ecosystem with randomly interacting species", *J. Phys. A* 22: 3447-3460.
Spagnolo, B. (1995) "Noise in nonlinear dynamical systems", *Accademia Peloritana dei Pericolanti*, vol. LXXII: 165-183.

ANALYZING THE MULTIFRACTAL STRUCTURE OF DNA NUCLEOTIDE SEQUENCES

J.M. GUTIERREZ[†], A. IGLESIAS[†], M.A RODRIGUEZ[‡],
[†]*Departamento de Matemática Aplicada,*
[‡]*Instituto de Física de Cantabria,*
Universidad de Cantabria, E-39005, Spain
E-mail: gutierjm@ccaix3.unican.es
and
J.D. BURGOS[†], C.M. ESTEVEZ[†], and P. MORENO[‡]
[†]*Laboratorio de Biotecnologia. Universidad INCCA,*
[‡] *Departamento de Biología. P. Universidad Javeriana,*
AE-78001, Colombia

ABSTRACT

This paper presents some preliminary studies on the multifractal structure of DNA nucleotide sequences and gives new insight into some recently appeared algorithms for characterizing the genetic information stored in these sequences. The basic idea is considering the sequence as a measure laying on the support of an Iterated Function System model. This measure is analyzed by using its multifractal spectrum, or the associated generalized dimensions. The multifractal spectrum contains all the statistical information related to subsequences of nucleotides within the DNA sequence, so it provides an appropriate tool for analyzing the underlying structure (repeated subsequences, correlations, etc.).

1. Introduction

During the last few years, numerous methods have been introduced to characterize and graphically represent the genetic information stored in DNA nucleotide sequences, the natural storage of the hereditary genetic information. The goal of these methods is generating representative patterns for certain sequences, or groups of sequences with the aim of obtaining some information about the underlying structure. Several standard techniques have been already applied to this problem: random walks (Berthelsen et al., 1992; Buldyrev et al., 1993), wavelets (Altaiski et al., 1996; Arneodo et al., 1996), and other techniques from nonlinear dynamics, such as the "chaos game algorithm" (Jeffrey, 1990). All these methods exploit the particular composition of DNA, as a sequence of four possible nucleotides which define the structure of amino acids to form proteins. Thus, DNA nucleotide sequences can be considered as a symbolic sequence of a four letter alphabet: adenine (A), cytosine (C), thymine (T), and guanine (G).

In particular, the chaos game representation uses the theory of Iterated Function Systems (Barnsley, 1990) to produce a fractal pattern associated with a given DNA sequence. An interesting property, from a biological point of view, is that some groups of organisms (protozoa, viruses, vertebrates, etc.) present common characteristic patterns when analyzed with this method. This gives some clues about the genetic similarity of organisms in the same evolution stage.

The basic idea of this method is using the sequence of basis to draw points in a square each of its four corners being associated with one of the four basis. As a result of this process, if we consider a $2^m \times 2^m$ grid in the unit square, each of the regions of the resulting pattern is associated with a different subsequence of basis $b_1 \ldots b_m$ (see (Jeffrey, 1990) for more details). Thus, the support tells us about the existence or not of such combinations. For example, Figure 1 shows the pattern obtained by applying the chaos game representation to the DNA sequence with GenBank entry ECOUW89 (Escherichia Coli), which contains 176195 bases (a 256×256 grid is considered). The white clusters in this figure are associated with subsequences $b_1 \ldots b_8$ which are not contained in the sequence.

However, besides of the pattern itself, the local densities of points existing in different regions of the pattern give a valuable information about the structure of the sequence, since the density of a region is related to the frequency the associated subsequence appears within the original DNA sequence. In this paper, we deal with the measure defined by the sequence of points on the pattern, as the natural model to retain information about these local densities (see Fig. 1).

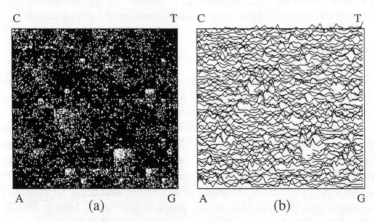

Figure 1: (a) Chaos game representation of the Escherichia Coli (176195 bases), and (b) associated multifractal measure.

2. The Proposed Method

Given a sequence of DNA nucleotides and considering the pattern generated by applying the chaos game algorithm, we extend this representation by introducing a measure on the pattern, or support, which represents the frequencies of the subsequences within the original DNA sequence. Therefore, by considering the sequence of points in the unit square provided by the above chaos game representation, we can define $\mu(S) = \#(S)/n$, where $\#(S)$ stands for the number of points of a total of n (the length of the sequence) laying on the subset S of the pattern P.

The structure of these measures can be conveniently analyzed using the multifractal formalism (see (Falconer, 1993) and (Gutiérrez et al., 1996)). This theory analyzes the scaling properties of subsets where the measure has a given density, say subsets $S(\alpha)$ formed by points x where $\mu(B(x,r)) \approx r^\alpha$ for small r (where $B(x,r)$ is the r-ball above x). That is, if we fix the size of the balls, the smaller the singularity associated with a point x, the larger the measure contained around the point. The main objective of the multifractal analysis is characterizing the structure associated with the sets $S(\alpha)$ as a function of the "density exponents" α (usually called singularities of the measure, due to some analogies with some physical phenomena). This information is given by the multifractal spectrum diagram α vs $f(\alpha)$, where $f(\alpha)$ is the Haussdorff dimension of the set $S(\alpha)$. Thus, the multifractal spectrum tells us how certain subsets of the pattern with the same local density scale when we increase the resolution of the representation (i.e., when we consider a longer grid on the unit square). An alternative representation of the multifractal spectrum is sometimes given by the generalized dimensions D_q, $q \in \mathbb{R}$ that are not related to dimensions, but to entropies of the sequence (Grassberger and Procaccia, 1983).

3. Some Results

Figure 2 shows both the multifractal spectrum and the generalized dimensions of the measure shown in Fig. 1(b). Small values of α indicate high-density regions of the support, whereas high singularity values are associated with regions with small density.

This figure also compares the multifractal properties of the original sequence with the multifractal properties of a control measure generated randomly and matching the original sequence in base contains (i.e. rearranging the sequence at random to destroy the correlation structure). Thus, by comparing both spectra we can obtain information about the structure contained in the DNA sequence, since the existence of singularities in the DNA sequence with values different

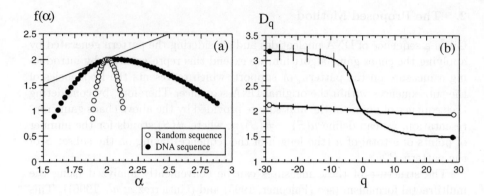

Figure 2: (a) The multifractal spectrum and (b) the generalized dimensions for both the DNA measure given in Figure 1(b), and an associated uncorrelated measure.

from those existing in the control sequence indicates a different distribution of subsequences between the control and the original sequences. For example, it can be seen that the DNA sequence shown in Fig. 1 contains a significant number of repeated subsequences (those associated with small singularity values).

Preliminary results also show that this new technique extends the results found by Jeffrey (Jeffrey, 1990) about the genetic similarities existing in certain groups of genes. Now, we can also conclude that some subsequences of nucleotides appear with similar frequencies in the DNA sequences of certain groups.

Moreover, there is a relationship between the multifractal spectrum and the correlation structure. This is an interesting result since, during the last years, an increasing attention has been focused on the development of efficient mechanisms to analyze the correlation structure of DNA sequences. This interest has been motivated by the controversy of the existence of long-range correlations in DNA sequences, and the differences between coding and noncoding DNA sequences for this kind of correlation (an extensive bibliography to this subject can be found in (Wentian, 1997)). Strongly related to this issue is the role of the noncoding regions in DNA sequences, whose functions (if any) still remain unclear. The multifractal technique introduced in this paper seems to be an appropriate tool for analyzing this problem. For example, the DNA sequence shown in Fig. 1 is mostly coding (over than 80 percent of the basis). Thus, although the fractal pattern in Fig. 1 seems to be quite uniform, the multifractal spectrum in Fig. 2 shows that this sequence contains an important correlation structure.

References

Altaiski, M., O. Mornev, and R. Polozov (1996) "Wavelet Analysis of DNA Sequences", *Genetic Analysis: Biomolecular Engineering*, 12, 165–168.

Arneodo, A., Y. d'Aubenton-Carafa, E. Bacry, P.V. Graves, J.F. Muzy, and C. Thermes (1996) "Wavelet Based Fractal Analysis of DNA Sequences", *Physica D*, 96, 291–320.

Barnsley, M.F. (1990) *Fractals Everywhere, 2nd edition*, (Academic Press, New York).

Berthelsen, C.L., J.A. Glazier, and M.H. Skolnick (1992) "Global Fractal Dimension of Human DNA Sequences Treated as Pseudorandom Walks", *Physical Review A*, 45, 8902–8913.

Buldyrev, V, A.L. Goldberger, S. Havlin, C-K. Peng, M. Simon, and H.E. Stanley (1993) "Generalized Levy-walk Model for DNA Nucleotide Sequences", *Physical Review E*, 47:6, 4514–4523.

Falconer, K. (1993) *Fractal Geometry* (John Wiley and Sons, Chichester).

Gutiérrez, J.M, A. Iglesias, and M.A. Rodríguez (1996) "A Multifractal Analysis of IFSP Invariant Measures with Application to Fractal Image Generation", *Fractals*, 4, 17–27.

Grassberger and Procaccia (1983) "Generalized Dimensions of Strange Attractors", *Physica D*, 9, 189–197.

Jeffrey, H.J. (1990) "Chaos Game Representation of Gene Structure", *Nucleic Acids Research*, 18:8, 2163–2175.

Wentian, L. (1997) "A Bibliography on Long-Range Correlation in DNA Sequences", "http://linkage.rockefeller.edu/wli/dna_corr/list.html"

CHAOTIC FIRING OF CORTICAL NEURONS UPON INHIBITORY PERIODIC STIMULATION

R. STOOP

Inst. für Neuroinformatik, ETHZ/UNIZH,
Gloriastr.32, CH-8006 Zürich, Switzerland
E-mail: ruedi@ini.phys.ethz.ch

K. SCHINDLER, and P. GOODMAN

Inst. für Neuroinformatik, ETHZ/UNIZH,
Gloriastr.32, CH-8006 Zürich, Switzerland

ABSTRACT

We investigate the response of regularly spiking rat cortical cells upon periodic stimulation. When inhibitory stimulation is applied, for specific ratios between stimulation and self oscillation frequency the resulting spiking pattern is proven to be chaotic. This ability of cortical neurons may be of importance for the desynchronization of a network of neurons.

Although considerable progress has been achieved in the past, the way the brain works is still far from being understood (Johnston and Wu, 1995). Understanding the brain is intrinsically connected with questions like how information is stored and propagated. Like any computational device, the brain is built up from computational elements. Clearly, the connection of such cells leads to a degree of complexity which goes beyond the complexity of the computational elements. However, it is not unreasonable to expect that the behavior of the whole brain can be related to the behavior of single elements. Recently, the question attained much interest whether brain activity can be chaotic (Schiff et al., 1994). In our contribution we shall prove that single cortical cells may respond with chaotic firing patterns to periodic inhibitory stimulation. In order to extrapolate from this fact on the behavior of a network of such cells, a coupled map lattice approach is necessary (Bunimovich and Sinai, 1988; Bunimovich, 1995; Bunimovich and Ventakagiri, 1997).

It is well known that single spiking neurons can be described as electronic oscillators (Douglas and Martin, 1992). Unfortunately, the integration of the resulting differential equations amounts to a rather time-consuming job, with a large set of constants to be adjusted. The coupling of such cells on lattices then results in a problem which makes the use of super computers unavoidable. From the point of view of nonlinear dynamics, a periodically spiking neuron is represented by a limit cycle solution. The idea to convert the differential equation system into a discrete map by means of a Poincaré section then leads to a considerable simplification of the problem. Incoming information in the form of stimuli reduces in this picture to a perturbation of the limit cycle. In a similar context (that of the embryonic chicken heart cell beating) this point of view has been adopted with considerable success some time ago by Glass and Mackey (Glass et al., 1984; Glass and Mackey, 1988). In our experiment, we took slices of rat brains in an in vitro preparation (for details of the preparation and more biological aspects see Schindler et al., 1997). We performed two sets of experiments. We first started by directly introducing a sharp micro-pipette into the cell. The limit cycle behavior was triggered by the application of a constant DC current to the cell. In order to obtain the perturbation of the limit cycle, this current was modified at the chosen stimulation times by an addition (excitatory stimulation) or by a subtraction (inhibitory stimulation) of a short-time current pulse of the duration of about $5ms$. Excitatory stimulation leads to a quicker occurrence of the next spike, whereas inhibitory stimulation has the general tendency of postponing the occurrence of a spiking for some time. In order to obtain a more biologically motivated type of stimulation, in a second set of experiments we used a bipolar electrode to stimulate the cell by means of exciting nerve fibers leading to this neuron (this then produces a *synaptic input*). For inhibitory synaptic stimulation we had to exclude excitatory stimulation by using pharmacological blockers (for details see Schindler et al., 1997). The response of the neuron upon an incoming perturbation results in a modified phase at which the next spike appears. In the experiment this phase can, according to Fig. 1, be determined as follows (Glass et al., 1984; Glass and Mackey, 1988)

$$T + t_2 = t_1 + T_s, \tag{1}$$

where T_s is the time between successive perturbations, T is the perturbed cycle length, t_1 is the time after spiking at which the perturbation was applied, and t_2 is the time after spike at which the next perturbation will appear (t_2 can be viewed as encoding the length of the perturbed cycle). Expressing this relation in terms of phases is achieved by first subtracting T on both sides of the equation and then dividing the equation by T_0, the cycle time of the unperturbed limit cycle. This leads to the equation

$$P: \Phi_2 = \Phi_1 + \Omega - T/T_0, \tag{2}$$

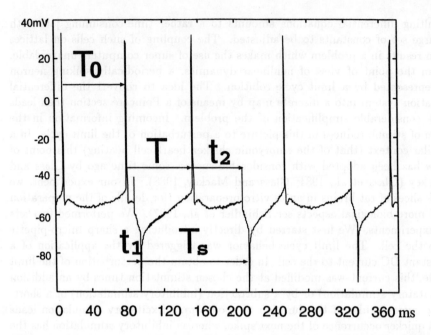

Figure 1: Example of an experimental spike train, where the characteristic quantities T, T_0, t_1, t_2 and T_s are indicated.

where Ω is the phase shift between the periodic limit cycle and the periodic perturbation. This equation, which has the form of a circle map (Cornfeld et al., 1982), can be interpreted as a Poincaré return map. The reaction of the cell upon the stimulation is essentially contained in the last right-hand-side contribution in this equation. This function can be called the *phase-response* function (Glass and Mackey, 1988) $g: t_1/T_0 \rightarrow T/T_0$. For the neuron stimulated by inhibition, typically a phase response function as shown in Fig. 2 is obtained, where the measured experimental data are shown by points and the interpolated function is shown as a line. The resulting Poincaré map can then be iterated, in order to obtain a prediction of the phases which should be observed upon a continued perturbation of the system. Periodic behavior, for example, is identified by a discrete number of possible phases. It is worth emphasizing that the resulting phases are not entirely determined by the phase response map g, but also depend on the value of the phase shift Ω which has a strong influence on the grammatical structure of the system. An investigation of the produced phases in dependence of Ω leads to a bifurcation diagram. The typical bifurcation diagram for our experiment in the case of inhibitory stimulation is displayed in Fig. 3.

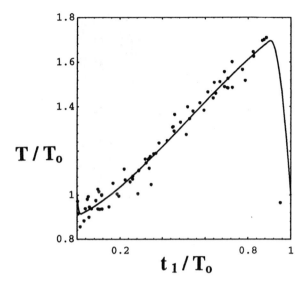

Figure 2: Phase response curve $g : t_1 \to T$ (see Fig. 1). This function expresses how the limit cycle responds to the perturbation given at the time t_1 after the last spike. When inserted in the circle- type map (c.f. Eq. 2), it determines at which new phase the next perturbation occurs, given the phase of the old perturbation.

Starting with periodic behavior at small phase shifts, soon bands of phases arise which indicate irregular response of the neuron upon the perturbation. In order to investigate the nature of these irregularities, we calculated the Lyapunov exponents which correspond to the bifurcation diagram. The result is shown in Fig. 4. As is well known, positive Lyapunov exponents indicate chaotic behavior, negative exponents indicate stable behavior (Stoop and Meier, 1988; Peinke et al., 1992). Clearly, areas where positive exponents arise appear in the diagram. Their existence proves that a chaotic response of the neuron upon periodic inhibitory stimulation is possible, for suitable choices of Ω. Our experimental data indicates that for excitatory stimulation, chaotic response is not possible.

We compared the results for inhibitory stimulation from numerous cells ($n >$ 20) and obtained qualitatively identical phase response curves, for both physical and synaptic stimulation. Moreover, even when different classes of maps were used in order to approximate the experimental phase response functions, they all led to qualitatively identical bifurcation diagrams, showing an astonishing stabil-

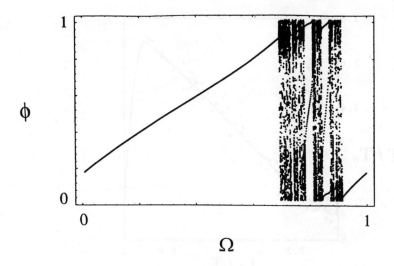

Figure 3: Bifurcation diagram produced by the inhibition-stimulated neuron. When the periodic perturbation with a fixed phase-shift Ω is performed (where Ω is the value on the x-axis), the points in the diagram on the parallel to y-axis indicate the phases which emerge. Regular firing corresponds to a discrete set of phases. In this plot, Ω is called the bifurcation parameter.

ity of the observed response. Excitatory stimulation also led to stable nontrivial bifurcation diagrams, where, however, no chaotic region was observed, because in this case the corresponding Poincaré map is always invertible (again the results of the two kinds of stimulation agree perfectly). Of course, the bifurcation diagram also depends on the perturbation strength. Our experimental evidence indicates that within a broad range of biologically meaningful perturbation strengths k this dependence can be modeled as

$$g(k,x) = (g_{ko}(x) - 1)k + 1, \tag{3}$$

where $x := t_1/T_0$, $g(k,x)$ is the phase response function at (arbitrary) perturbation strength k and g_{ko} denotes the phase response function obtained for a chosen perturbation strength ko. We observed that within broad parameter range of the stimulation strength the qualitative features of the bifurcation diagram remained unchanged. For the explicit discussion of this dependence we refer to Stoop,

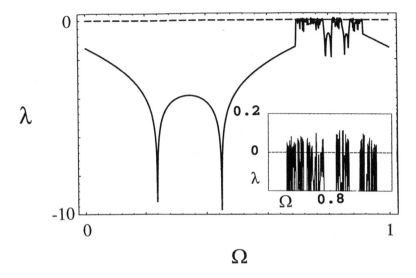

Figure 4: Lyapunov exponents diagram obtained upon stimulation by inhibition. Bands with positive Lyapunov exponents prove the existence of chaotic firing patterns at the corresponding values of the frequency shift Ω.

work in preparation. As a general tendency we mention that with increasing perturbation strength, the chaotic bands get narrower and are shifted towards high Ω-values. Furthermore, it is of importance to compare the predictions made by the bifurcation diagram with the results obtained from experiments of continued perturbations to prove that under the given conditions the assumption of a stable enough limit cycle still is valid. To this end, we stimulated the neuron periodically with inhibition at a fixed value of Ω, classified the obtained set of phase shifts as periodic or as chaotic and then compared these findings with the predictions made by the bifurcation diagram. We observed a good agreement between prediction and experiment. We finally investigated the influence of additive Gaussian distributed white noise on the bifurcation diagram and compared the results with direct experimental observation. It is found that within the range of noise observed in the experiment, the qualitative features of the bifurcation diagram remain unchanged.

Our conclusion is that whenever inhibitory stimulation is applied, neurons can respond chaotically, which is not the case for excitatory stimulation. In networks of neurons, as a general tendency, excitatory stimulation enhances synchronized activity. For the network it may be of value to have an efficient means by which this synchrony can be broken. We propose that one possible mechanism to achieve this task is to drive the periodically spiking neurons by means of inhibitory stimulation into a regime where they start to fire chaotically, breaking in this way the synchrony of the system. More detailed consequences of the observed firing properties can be worked by considering networks of cells exposed to inhibitory and excitatory stimulation. Present results include spatio-temporal chaos and traveling patterns (Stoop, work in preparation). This work was partially supported by the Swiss National Science and the Maurice E.Müller Foundation.

References

Bunimovich, L.A. and Ya.G. Sinai (1988), "Spacetime chaos in coupled map lattices", Nonlinearity **1**, 491-516.

Bunimovich, L.A. (1995), "Coupled map lattices: One step forward and two steps back", Physica D **86**, 248-255.

Bunimovich, L.A. and S. Ventakagiri (1997), in *Exact Approaches to Irregular Systems*, R. Stoop and G. Radons, eds., Phys. Rep., to appear.

Cornfeld, I.P., S.V. Fomin and Ya.G. Sinai (1982), *Ergodic Theory*, Berlin: Springer.

Douglas, R.J. and K.A.C. Martin(1993), "Exploring Cortical Microcircuits", in *Single Neuron Computation*,
T. McKenna, J. Davies, and S.F. Zornetzer eds, Boston: Academic Press.

Glass L., M. Guevara, J. Belair, and A. Shrier (1984), "Global bifurcations of a periodically forced oscillator", Phys. Rev. A **29**, 1348-1357.

Glass L. and M. Mackey (1988), *From Clocks to Chaos*, Princeton: Princeton University Press.

Johnston, D. and S.M.-S. Wu (1995), *Foundations of Cellular Neurophysiology*, Cambridge, Mass.: MIT-Press.

Peinke, J., J. Parisi, O.E. Roessler, and R. Stoop (1992), *Encounter with Chaos*, Berlin: Springer.

Schiff, S.J., K. Jerger, D.H. Dung, T. Chang, M.L. Spano, and W.L. Ditto (1994), "Controlling Chaos in the Brain", Nature **370**, 615-620.

Schindler, K., C. Bernasconi, R. Stoop, P. Goodman, and R.Douglas, work to appear (1997).

Schuster, H.G. (1989), *Deterministic Chaos*, 2nd ed., Weinheim: VCH.

Stoop, R., work in preparation.

Stoop, R. and P.F. Meier (1988), "Evaluation of Lyapunov exponents and scaling functions from time series", J. Opt. Soc. Am. B **5**, 1037-1045.

PARTICIPANTS

List of the participants

Ladislav ANDREY
Institute of Computer Science
Pod Vodarenskou Vezi 2
18207-Prague 8 (Czech Republic)
Tel. +420-2-66052085
Fax: +420-2-8585789
E-mail: andre@uivt.cas.cz

Tito ARECCHI
Istituto Nazionale di Ottica
Largo E. Fermi 6
50125-Firenze (Italy)
Tel. +39-55-23081
Fax: +39-55-2337755
E-mail: arecchi@firefox.ino.it

Rita BALOCCHI
Istituto di Fisiologia Clinica, CNR
Via Trieste 41
56126-Pisa (Italy)
Tel. +39-50-502771
Fax: +39-50-589038
E-mail: balocchi@nsifc.ifc.pi.cnr.it

Maria BARBI
Dipartimento di Fisica
Università di Firenze
Address home: via G. di Simone 12
56120-Pisa (Italy)
Tel. (home) +39-50-570886
E-mail: maria@avanzi.de.unifi.it

Michele BARBI
Istituto di Biofisica, CNR
Via S. Lorenzo 26
56127-Pisa (Italy)
Tel. +39-50-513111
Fax: +39-50-553501
E-mail: barbi@ib.pi.cnr.it

Gianfranco BASTI
Facoltà di Filosofia
Università Pontificia Gregoriana
Piazza della Pilotta 4
00187-Roma (Italy)
Tel. +39-6-4742529
Fax: +39-6-4742529
E-mail: basti@roma2.infn.it

Stefano BOCCALETTI
Istituto Nazionale di Ottica
Largo E. Fermi 6
50125-Firenze (Italy)
Tel. +39-55-23081
Fax: +39-55-2337755
E-mail: stefano@firefox.ino.it

Luca BONCI
Dipartimento di Fisica
Università di Pisa
Piazza Torricelli 2
56127-Pisa (Italy)
Tel. +39-50-911265
Fax: +39-50-48277
E-mail: bonci@ipifidpt.difi.unipi.it

Aldo CASALEGGIO
Istituto per i Circuiti Elettronici CNR
Via De Marini 6
16149-Genova (Italy)
Tel. +39-10-6475218
Fax: +39-10-6475200
E-mail: casaleggio@ice.ge.cnr.it

Roberto CHIGNOLA
Istituto di Immunologia e Malattie
Infettive, Università di Verona
Policlinico Borgo Roma
37100-Verona (Italy)
Tel. +39-45-8074256
Fax: +39-45-580900
E-mail: chignola@borgoroma.univr.it

Santi CHILLEMI
Istituto di Biofisica, CNR
Via S. Lorenzo 26
56127-Pisa (Italy)
Tel. +39-50-513111
Fax: +39-50-553501
E-mail: chillemi@ib.pi.cnr.it

Vladimir CHINAROV
Sci. Res. Center "Vidhuk"
Vladimirskaya Str. 61-b
252033-Kiev (Ukraine)
Tel. +38-44-220-1403
Fax: +38-44-227-4482
E-mail:chinarov@vidguk.freenet.kiev.ua

Paolo CHIOCCON
Facoltà di Scienze Statistiche
Università di Bologna
Via delle Belle Arti 41
40126-Bologna (Italy)
Tel. +39-51-258219
Fax: +39-51-227997
E-mail: chioccon@unive.it

Simona COCCO
Dipartimento di Fisica
Università di Roma "La Sapienza"
Piazzale A. Moro 2
00185-Roma (Italy)
Tel. +39-6-8086462
Fax: +39-6-4463158
E-mail: COCCO@roma1.infn.it

James J. COLLINS
Dept. of Biomedical Engineering
Boston University
44 Cummington Str.
MA 02215-Boston (USA)
Tel. +1-617-353-9635
Fax: +1-617-353-5737
E-mail: collins@enga.bu.edu

Antonio COTUGNO
Istituto di Cibernetica, CNR
Via Toiano 6
80072-Arco Felice-Napoli (Italy)
Tel. +39-81-8534-113/131
Fax: +39-81-5267654
E-mail: neuros@mail.irtemp.na.cnr.it

Angelo DI GARBO
Istituto di Biofisica, CNR
Via S. Lorenzo 26
56127-Pisa (Italy)
Tel. +39-50-513111
Fax: +39-50-553501
E-mail: digarbo@ib.pi.cnr.it

Shulamit EYAL
Tel-Aviv University
Center of Medical Physics
School of Physics and Astronomy
69978-Tel-Aviv (Israel)
Tel. +972-3-6408669
Fax: +972-3-6406237
E-mail: shuli@post.tau.ac.il

Roberto FIORAVANTI
Ist. di Cibernetica e Biofisica, CNR
Via De Marini 6
16149-Genova (Italy)
Tel. +39-10-6475564
Fax: +39-10-6475500
E-mail: roberto@barolo.icb.ge.cnr.it

Leone FRONZONI
Dipartimento di Fisica
Università di Pisa
P.zza Torricelli, 2
56127-Pisa (Italy)
Tel. +39-50-911235
Fax: +39-50-48277
E-mail: fronzoni@ipifidpt.difi.unipi.it

Ilse Christine GEBESHUBER
TU-BioMed, University of Technology
Wiedner Hauptstrasse 8-10/1145
A-1040-Wien (Austria)
Tel. +43-1-58801-4139
Fax: +43-1-5862959
E-mail: igebes@fbma.tuwien.ac.at

Philip H. GOODMAN
University of Nevada, Reno ETH/U
Hochstrasse 58
8044-Zürich (Switzerland)
Tel. +41-1-251-3578
Fax: +41-1-257-6983
E-mail: goodman@ini.phys.ethz.ch

Celso GREBOGI
Laboratory for Plasma Research
Inst. for Physical Science & Technology
University of Mariland
MD 20742-College Park (USA)
Tel. +1-301-405-5021
Fax: +1-301-405-1678
E-mail: grebogi@chaos.umd.edu

Arturas GRIGALIUNAS
Kaunas Medical Academy
Eivenlu 4
3007-Kaunas (Lithuania)
Tel. +370-7732166
Fax: +370-7796498
E-mail: Arturas.Grigaliunas@gim.ktu.it

Jose Manuel GUTIERREZ
Dept. de Matematica Aplicada y
Ciensas de la Computacion
Universidad de Cantabria
Avda. de los Castros s/n
39005-Santander (Spain)
Tel. +34-42-201723
Fax: +34-42-201703
E-mail: gutierjm@ccaix3.unican.es

Arun V. HOLDEN
Dept. of Physiology
University of Leeds
The Worsley Medical and Dental Bldg.
LS2-9NQ-Leeds (UK)
Tel. +44-113-233-4251
Fax: +44-113-233-4228
E-mail: arun@www.cbiol.leeds.ac.uk

Peter Foster HUTTNER
Institute for Experimental Physics
University of Wien
Boltzmanngasse 5
A-1090-Wien (Austria)
Tel. +43-1-5266345 (private)
E-mail: huettner@acpx.exp.univie.ac.at

Juergen KURTHS
Center for Nonlinear Dynamics and
Max Planck Institute
University of Potsdam
D-14195-Potsdam (Germany)
Tel. +49-331-977-1611
Fax: +49-331-977-1142
E-mail:jkurths@agnld.uni-potsdam.de

Rocco MAMMOLITI
Dipartimento di Ingegneria Informatica
Università di Pisa
Via Diotisalvi 2
56126-Pisa (Italy)
Tel. +39-50-583304
Fax: +39-50-568522
E-mail: roccom@po.ifc.pi.cnr.it

Roberto MARANGONI
Univ. di Pisa and Ist. di Biofisica, CNR
Via S. Lorenzo 26
56127-Pisa (Italy)
Tel. +39-50-513111
Fax: +39-50-553501
E-mail: roberto@ib.pi.cnr.it

Michael MENZINGER
Dep. of Chemistry
University of Toronto
Ont. M5S.3H6-Toronto (Canada)
Tel. +1-416-978-6158
Fax: +1-416-978-6158
E-mail: mmenzing@alchemy.chem.utoronto.ca

Anna MONTAGNINI
Istituto di Biofisica, CNR
Via S. Lorenzo 26
56127-Pisa (Italy)
Tel. +39-50-513111
Fax: +39-50-553501
E-mail: annam@ib.pi.cnr.it

Frank MOSS
Center for Neurodynamics
Dept. of Physics & Astronomy
University of Missouri at St. Louis
MO 63121-St. Louis (USA)
Tel. +1-314-516-6150
Fax: +1-314-516-6152
E-mail: mossf@umslvma.umsl.edu

Carlo MUSIO
Istituto di Cibernetica, CNR
Via Toiano 6
80072-Arco Felice-Napoli (Italy)
Tel. +39-81-8534-113/131
Fax: +39-81-5267654
E-mail: camus@biocib.cib.na.cnr.it

Claudia NERI
Istituto di Biofisica, CNR
Via S. Lorenzo 26
56127-Pisa (Italy)
Tel. +39-50-513111
Fax: +39-50-553501
E-mail: neri@ib.pi.cnr.it

Taishin NOMURA
Dept. of Biophysical Engineering
Grad. School of Engineering Science
Osaka University
Toyonaka, Osaka 560 (Japan)
Tel. +81-6-850-6533
Fax: +81-6-850-6557
E-mail: taishin@bpe.es.osaka-u.ac.jp

Johnny T. OTTESEN
Dept. of Mathematics and Physics
Roskilde University
DK-4000-Roskilde (Denmark)
Tel. +45-46757781 (ext 2298)
Fax: +4546755065
E-mail: Johnny@mmf.ruc.dk

Khashayar PAKDAMAN
Dept. of Biophysical Engineering
Osaka University
Toyonaka, Osaka 560 (Japan)
Tel. +81-6-850-6533
Fax: +81-6-850-6557
E-mail: pakdaman@bpe.es.osaka-u.ac.jp

Antonio L. PERRONE
Dip. di Fisica
Università "Tor Vergata"
Via della Ricerca Scientifica 1
00133 Roma (Italy)
Tel. +39-6-4742529
Fax: +39-6-4742529
E-mail: perrone@roma2.infn.it

Donatella PETRACCHI
Istituto di Biofisica, CNR
Via S. Lorenzo 26
56127-Pisa (Italy)
Tel. +39-50-513111
Fax: +39-50-553501
E-mail: petracch@ib.pi.cnr.it

Claudia PETRONGOLO
Istituto di Biofisica, CNR
Via S. Lorenzo 26
56127-Pisa (Italy)
Tel. +39-50-513111
Fax: +39-50-553501
E-mail: claudia@ib.pi.cnr.it

Gian Domenico PINNA
Servizio di Bioingegneria
Fondazione S. Maugeri
Centro Medico di Riabilitazione
Via per Montescano
27040-Pavia (Italy)
Tel. +39-385-247256
Fax: +39-385-61386
E-mail: bing@fsm.it

Alberto PORTA
Dipartimento di Bioingegneria
Politecnico di Milano
Via Golgi 39
20133-Milano (Italy)
Tel. +39-2-2399-3302
Fax: +39-2-2399-3360
E-mail: porta@cdc8g5.cdc.polimi.it

Silvia SANTILLO
Istituto di Cibernetica, CNR
Via Toiano 6
80072-Arco Felice-Napoli (Italy)
Tel. +39-81-8534-113/131
Fax: +39-81-5267654
E-mail: neuros@mail.irtemp.na.cnr.it

Tim SAUER
George Mason University
Dept. of Mathematical Science
22030-4444 Fairfax, Virginia (USA)
Tel. +1-703-993-1471
Fax: +1-703-993-1491
E-mail: tsauer@osf1.gmu.edu

Steven J. SCHIFF
Center for Neuroscience
Children's Research Institute
111 Michigan Ave., NW
DC 20010-Washington (USA)
Tel. +1-202-884-3755
Fax: +1-202-884-6510
E-mail: sschiff@gwis2.circ.gwu.edu

Kaspar SCHINDLER
Institute for Neuroscience, ETH
University of Zürich
Gloriastr. 32
8006 Zürich (Switzerland)
Tel. +41-1-2572617
Fax: +41-1-2576983
E-mail: kaspar@ini.phys.ethz.ch

Agata SCORDINO
Istituto di Fisica, Facoltà d'Ingegneria
Università di Catania
Viale A. Doria 6
95125-Catania (Italy)
Tel. +39-95-256432
Fax: +39-95-256443
E-mail: demone@if.ing.unict.it

Maria SIGNORINI
Dipartimento di Bioingegneria
Politecnico di Milano
Piazza Leonardo da Vinci 32
20133-Milano (Italy)
Tel. +39-2-23993342
Fax: +39-2-23993360
E-mail: signorin@elet.polimit.it

Enrico SIMONOTTO
Dipartimento di Fisica
via Dodecaneso 33
16122-Genova (Italy)
Tel. +39-10-353-6314
Fax: +39-10-311066
E-mail: simon@ge.infm.it

Bernardo SPAGNOLO
Dipartimento di Energetica ed
Applicazioni di Fisica
Università di Palermo
Viale delle Scienze
90128-Palermo (Italy)
Tel. +39-91-236317
Fax: +39-91-236306
E-mail: spagnolo@ipacuc.it

Francesca SPANO'
Dipartimento di Fisica
via Dodecaneso 33
16122-Genova (Italy)
Tel. +39-10-353-6314
Fax: +39-10-311066
E-mail: spano@ge.infm.it

Mark L. SPANO
NSWC, Carderock Division, Code 684
Bldg. 19-A252
MD 20084-5000-Bethesda (USA)
Tel. +1-301-227-4466
Fax: +1-301-227-4449
E-mail: mark@chaos.dt.navy.mil

Nico STOLLENWERK
Research Center Jülich, MOD
D-52425-Jülich (Germany)
Tel. +49-2461-61-2982
Fax: +49-2461-61-2983
E-mail: N.Stollenwerk@fz-juelich.de

Ruedi STOOP
Institute for Neuroscience, ETH
University of Zürich
Gloriastr. 32
8006 Zürich (Switzerland)
Tel. +41-1-2572617
Fax: +41-1-2576983
E-mail: Stoop@msp.chem.ethz.ch

Cloe TADDEI-FERRETTI
Istituto di Cibernetica, CNR
Via Toiano 6
80072-Arco Felice-Napoli (Italy)
Tel. +39-81-8534-113/131
Fax: +39-81-5267654
E-mail: neuros@mail.irtemp.na.cnr.it

Gina TOCCHINI
Istituto di Biofisica, CNR
Via S. Lorenzo 26
56127-Pisa (Italy)
Tel. +39-50-513111
Fax: +39-50-553501
E-mail: ginetta@ib.pi.cnr.it